T0241794

MATHEMATICS IN INDUSTRY **20**

More information about this series at http://www.springer.com/series/4650

Peter Benner

Editor

System Reduction for Nanoscale IC Design

 Springer

Editor
Peter Benner
Max Planck Institute for Dynamics
    of Complex Technical Systems
Magdeburg, Germany

ISSN 1612-3956   ISSN 2198-3283   (electronic)
Mathematics in Industry
ISBN 978-3-319-79154-8   ISBN 978-3-319-07236-4   (eBook)
DOI 10.1007/978-3-319-07236-4

Mathematics Subject Classification (2010): 94-02, 65L80, 94C05

Printed on acid-free paper

This Springer imprint is published by Springer Nature
The registered company is Springer International Publishing AG
The registered company address is: Gewerbestrasse 11, 6330 Cham, Switzerland

Preface

The ongoing miniaturization of devices like transistors used in integrated circuits (ICs) has led to feature sizes on the nanoscale. The Intel Core 2 (Yorkfield), first presented in 2007, was produced using 45 nm technology. Recently, production has reached 14 nm processes, e.g., in the Intel Broadwell, Skylake, and Kaby Lake microprocessors. Although the main principles in IC design and production are those of microelectronics, nowadays, one therefore speaks of *nanoelectronics*.

With miniaturization now reaching double-digit nanometer length scales and the huge number of semiconductor devices employed, which result in a correspondingly significant rise in integration density, the influence of the wiring and supply networks (interconnect and power grids) on the physical behavior of an IC can no longer be neglected and must be modeled with the help of dedicated network equations in the case of computer simulations. Furthermore, critical semiconductor devices can often no longer be modeled by substitute schematics as done in the past, using, e.g., the Partial Element Equivalent Circuit (PEEC) method. Instead, complex mathematical models are used, e.g., the drift-diffusion model. In addition to shortened production cycles, these developments in the design of new nano-electronic ICs now increasingly pose challenges in computer simulations regarding the optimization and verification of layouts. Even in the development stage, it has become indispensable to test all crucial circuit properties numerically. Thus, the field of *computational nanoelectronics* has emerged.

The complexity of the mathematical models investigated in computational nanoelectronics is enormous: small parts of an IC design alone may require millions of linear and nonlinear differential-algebraic equations for accurate modeling, allowing the prediction of its behavior in practice. Thus, the full simulation of an IC design requires tremendous computational resources, which are often unavailable to microprocessor designers. In short, one could justifiably claim that *the performance of today's computers is too low to simulate their successors!*—a statement that has been true for the last few decades and is debatably still valid today. Thus, the dimension reduction of the mathematical systems involved has become crucial over the past two decades and is one of the key technologies in computational nanoelectronics.

The dimension or model reduction at the system level, or *system reduction* for short, is mostly done by mathematical algorithms, which produce a much smaller (often by factors of 100 up to 10,000) model that reproduces the system's response to a signal up to a prescribed level of accuracy. The topic of *system reduction in computational nanoelectronics* is the focus of this book. The articles gathered here are based on the final reports for the network *System Reduction in Nanoscale IC Design (SyreNe)*, supported by Germany's Federal Ministry of Education and Research (BMBF) as part of its Mathematics for Innovations in Industry and Services program. It was funded between July 1, 2007, and December 31, 2010 (see syrene.org for a detailed description) and continued under the name *Model Reduction for Fast Simulation of new Semiconductor Structures for Nanotechnology and Microsystems Technology (MoreSim4Nano)* within the same BMBF funding scheme from October 1, 2010, until March 31, 2014 (see moresim4nano.org).

The goal of both research networks was to develop and compare methods for system reduction in the design of high-dimensional nanoelectronic ICs and to test the resulting mathematical algorithms in the process chain of actual semiconductor development at industrial partners. Generally speaking, two complementary approaches were pursued: the reduction of the nanoelectronic system as a whole (subcircuit model coupled to device equation) by means of a global method and the creation of reduced order models for individual devices and large linear subcircuits which are linked to a single reduced system. New methods for nonlinear model reduction and for the reduction of power grid models were developed to achieve this.

The book consists of five chapters, introducing novel concepts for the different aspects of model reduction of circuit and device models. These include:

- Model reduction for device models coupled to circuit equations in Chap. 1 by Hinze, Kunkel, Matthes, and Vierling
- Structure-exploiting model reduction for linear and nonlinear differential-algebraic equations arising in circuit simulation in Chap. 2 by Stykel and Steinbrecher
- The reduced representation of power grid models in Chap. 3 by Benner and Schneider
- Numeric-symbolic reduction methods for generating parameterized models of nanoelectronic systems in Chap. 4 by Schmidt, Hauser, and Lang
- Dedicated solvers for the generalized Lyapunov equations arising in balanced truncation based model reduction methods for circuit equations in Chap. 5 by Bollhöfer and Eppler

The individual chapters describe the new algorithmic developments in the respective research areas over the course of the project. They can be read independently of each other and provide a tutorial perspective on the respective aspects of System Reduction in Nanoscale IC Design related to the sub-projects within SyreNe. The aim is to comprehensively summarize the latest research results, mostly published in dedicated journal articles, and to present a number of new aspects never before published. The chapters can serve as reference works, but should also inspire future research in computational nanoelectronics.

I would like to take this opportunity to express my gratitude to the project partners Matthias Bollhöfer and Heike Faßbender (both from the TU Braunschweig), Michael Hinze (University of Hamburg), Patrick Lang (formerly the Fraunhofer-Institut für Techno- und Wirtschaftsmathematik (ITWM), Kaiserslautern), Tatjana Stykel (at the TU Berlin during the project and now at the University of Augsburg), Carsten Neff (NEC Europe Ltd. back then), Carsten Hammer (formerly Qimonda AG and then Infineon Technologies AG), and Peter Rotter (Infineon Technologies AG back then). Only their cooperation within SyreNe and their valued work in the various projects made this book possible.

Furthermore, I would like to particularly thank André Schneider, who helped in countless ways during the preparation of this book. This includes the LaTeX setup as well as indexing and resolving many conflicts in the bibliographies. Without his help, I most likely never would have finished this project. My thanks also go to Ruth Allewelt and Martin Peters of Springer-Verlag, who were very supportive and encouraging throughout this project. Their endless patience throughout the many delays in the final phases of preparing the book is greatly appreciated!

Magdeburg, Germany   Peter Benner
December 2016

Contents

Chapter 1
Model Order Reduction of Integrated Circuits in Electrical Networks

Michael Hinze, Martin Kunkel, Ulrich Matthes, and Morten Vierling

Abstract We consider integrated circuits with semiconductors modeled by modified nodal analysis and drift-diffusion equations. The drift-diffusion equations are discretized in space using mixed finite element method. This discretization yields a high-dimensional differential-algebraic equation. Balancing-related model reduction is used to reduce the dimension of the decoupled linear network equations, while the semidiscretized semiconductor models are reduced using proper orthogonal decomposition. We among other things show that this approach delivers reduced-order models which depend on the location of the semiconductor in the network. Since the computational complexity of the reduced-order models through the nonlinearity of the drift-diffusion equations still depend on the number of variables of the full model, we apply the discrete empirical interpolation method to further reduce the computational complexity. We provide numerical comparisons which demonstrate the performance of the presented model reduction approach. We compare reduced and fine models and give numerical results for a basic network with one diode. Furthermore we discuss residual based sampling to construct POD models which are valid over certain parameter ranges.

1.1 Introduction

Computer simulations play a significant role in design and production of very large integrated circuits or chips that have nowadays hundreds of millions of semiconductor devices placed on several layers and interconnected by wires. Decreasing

M. Hinze (✉) • U. Matthes • M. Vierling
Department of Mathematics, University of Hamburg, Bundesstraße 55, 20146 Hamburg, Germany
e-mail: michael.hinze@uni-hamburg.de; ulrich.matthes@math.uni-hamburg.de;
morten.vierling@uni-hamburg.de

M. Kunkel
Fakultät für Luft- und Raumfahrttechnik, Universität der Bundeswehr München,
Werner-Heisenberg-Weg 39, 85577 Neubiberg, Germany
e-mail: ich@martinkunkel.de

© Springer International Publishing AG 2017
P. Benner (ed.), *System Reduction for Nanoscale IC Design*,
Mathematics in Industry 20, DOI 10.1007/978-3-319-07236-4_1

physical size, increasing packing density, and increasing operating frequencies necessitate the development of new models reflecting the complex continuous processes in semiconductors and the high-frequency electromagnetic coupling in more detail. Such models include complex coupled partial differential equation (PDE) systems where spatial discretization leads to high-dimensional ordinary differential equation (ODE) or differential-algebraic equation (DAE) systems which require unacceptably high simulation times. In this context model order reduction (MOR) is of great importance. In the present work we as a first step towards model order reduction of complex coupled systems consider electrical circuits with semiconductors modeled by drift-diffusion (DD) equations as proposed in e.g. [46, 52]. Our general idea of model reduction of this system consists in approximating this system by a much smaller model that captures the input-output behavior of the original system to a required accuracy and also preserves essential physical properties. For circuit equations, passivity is the most important property to be preserved in the reduced-order model.

For linear dynamical systems, many different model reduction approaches have been developed over the last 30 years, see [6, 42] for recent collection books on this topic. Krylov subspace based methods such as PRIMA [32] and SPRIM [15, 16] are the most used passivity-preserving model reduction techniques in circuit simulation. A drawback of these methods is the *ad hoc* choice of interpolation points that strongly influence the approximation quality. Recently, an optimal point selection strategy based on tangential interpolation has been proposed in [3, 20] that provides an optimal $\mathbb{H}_2$-approximation.

An alternative approach for model reduction of linear systems is balanced truncation. In order to capture specific system properties, different balancing techniques have been developed for standard and generalized state space systems, see, e.g., [19, 31, 35, 37, 49]. In particular, passivity-preserving balanced truncation methods for electrical circuits (PABTEC) have been proposed in [38, 39, 51] that heavily exploit the topological structure of circuit equations. These methods are based on balancing the solution of projected Lyapunov or Riccati equations and provide computable error bounds.

Model reduction of nonlinear equation systems may be performed by a trajectory piece-wise linear approach [40] based on linearization, or proper orthogonal decomposition (POD) (see, e.g., [45]), which relies on snapshot calculations and is successfully applied in many different engineering fields including computational fluid dynamics and electronics [23, 29, 45, 48, 53]. A connection of POD to balanced truncation was established in [41, 54].

A POD-based model reduction approach for the nonlinear drift-diffusion equations has been presented in [25], and then extended in [23] to parameterized electrical networks using the greedy sampling proposed in [33]. An advantage of the POD approach is its high accuracy with only few model parameters. However, for its application to the drift-diffusion equations it was observed that the reduction of the problem dimension not necessarily implies the reduction of the simulation time. Therefore, several adaption techniques such as missing point estimation [4] and discrete empirical interpolation method (DEIM) [10, 11] have been developed to reduce the simulation cost for the reduced-order model.

In this paper, we review results of [23–27] related to model order reduction of coupled circuit-device systems consisting of the differential-algebraic equations modeling an electrical circuit and the nonlinear drift-diffusion equations describing the semiconductor devices. In a first step we show how proper orthogonal decomposition (POD) can be used to reduce the dimension of the semiconductor models. It among other things turns out, that the reduced model for a semiconductor depends on the position of the semiconductor in the network. We present numerical investigations from [25] for the reduction of a 4-diode rectifier network, which clearly indicate this fact. Furthermore, we apply the Discrete Empirical Interpolation Method (DEIM) of [10] for a further reduction of the nonlinearity, yielding a further reduction of the overall computational complexity. Moreover, we adapt to the present situation the Greedy sampling approach of [33] to construct POD models which are valid over certain parameter ranges. In a next step we combine the passivity-preserving balanced truncation method for electrical circuits (PABTEC) [38, 51] to reduce the dimension of the decoupled linear network equations with POD MOR for the semiconductor model. Finally, we present several numerical examples which demonstrate the performance of our approach.

1.2 Basic Models

In this section we combine mathematical models for electrical networks with mathematical models for semiconductors. Electrical networks can be efficiently modeled by a differential-algebraic equation (DAE) which is obtained from modified nodal analysis (MNA). Denoting by e the node potentials and by j_L and j_V the currents of inductive and voltage source branches, the DAE reads (see [18, 28, 52])

$$A_C \frac{d}{dt} q_C(A_C^\top e, t) + A_R g(A_R^\top e, t) + A_L j_L + A_V j_V = -A_I i_s(t), \qquad (1.1)$$

$$\frac{d}{dt} \phi_L(j_L, t) - A_L^\top e = 0, \qquad (1.2)$$

$$A_V^\top e = v_s(t). \qquad (1.3)$$

Here, the incidence matrix $A = [A_R, A_C, A_L, A_V, A_I] = (a_{ij})$ represents the network topology, e.g. at each non mass node i, $a_{ij} = 1$ if the branch j leaves node i and $a_{ij} = -1$ if the branch j enters node i and $a_{ij} = 0$ elsewhere. The indices R, C, L, V, I denote the capacitive, resistive, inductive, voltage source, and current source branches, respectively. The functions q_C, g and ϕ_L are continuously differentiable defining the voltage-current relations of the network components. The continuous functions v_s and i_s are the voltage and current sources.

Under the assumption that the Jacobians

$$D_C(e,t) := \frac{\partial q_C}{\partial e}(e,t), \quad D_G(e,t) := \frac{\partial g}{\partial e}(e,t), \quad D_L(j,t) := \frac{\partial \phi_L}{\partial j}(j,t)$$

are positive definite, analytical properties (e.g. the index) of DAE (1.1)–(1.3) are investigated in [14] and [13]. In linear networks, the matrices D_C, D_G and D_L are positive definite diagonal matrices with capacitances, conductivities and inductances on the diagonal.

Often semiconductors themselves are modeled by electrical networks. These models are stored in a library and are stamped into the surrounding network in order to create a complete model of the integrated circuit. Here we use a different approach which uses the transient drift-diffusion equations as a continuous model for semiconductors. Advantages are the higher accuracy of the model and fewer model parameters. On the other hand, numerical simulations are more expensive. For a comprehensive overview of the drift-diffusion equations we refer to [1, 2, 8, 30, 43]. Using the notation introduced there, we have the following system of partial differential equations for the electrostatic potential $\psi(t,x)$, the electron and hole concentrations $n(t,x)$ and $p(t,x)$ and the current densities $J_n(t,x)$ and $J_p(t,x)$:

$$\mathrm{div}(\varepsilon\,\mathrm{grad}\,\psi) = q(n - p - C),$$

$$-q\partial_t n + \mathrm{div}\,J_n = qR(n,p,J_n,J_p),$$

$$q\partial_t p + \mathrm{div}\,J_p = -qR(n,p,J_n,J_p),$$

$$J_n = \mu_n q(U_T\,\mathrm{grad}\,n - n\,\mathrm{grad}\,\psi),$$

$$J_p = \mu_p q(-U_T\,\mathrm{grad}\,p - p\,\mathrm{grad}\,\psi),$$

with $(t,x) \in [0,T] \times \Omega$ and $\Omega \subset \mathbb{R}^d (d = 1,\ldots,3)$. The nonlinear function R describes the rate of electron/hole recombination, q is the elementary charge, ε the dielectricity, μ_n and μ_p are the mobilities of electrons and holes. The temperature is assumed to be constant which leads to a constant thermal voltage U_T. The function C is the time independent doping profile. Note that we do not formulate into quasi-Fermi potentials since the additional non-linearities would imply higher simulation time for the reduced model. Further details are given in [23]. The analytical and numerical analysis of systems of this form is subject to current research, see [7, 17, 46, 52].

1.2.1 Coupling

In the present section we develop the complete coupled system for a network with n_s semiconductors. We will not specify an extra index for semiconductors, but we keep in mind that all semiconductor equations and coupling conditions need to be introduced for each semiconductor.

For the sake of simplicity we assume that to a semiconductor m semiconductor interfaces $\Gamma_{O,k} \subseteq \Gamma \subset \partial\Omega$, $k = 1,\ldots,m$ are associated, which are all Ohmic contacts, compare Fig. 1.2. The dielectricity ε shall be constant over the whole domain Ω. We focus on the Shockley-Read-Hall recombination

$$R(n,p) := \frac{np - n_i^2}{\tau_p(n + n_i) + \tau_n(p + n_i)}$$

which does not depend on the current densities. Herein, τ_n and τ_p are the average lifetimes of electrons and holes, and n_i is the constant intrinsic concentration which satisfy $n_i^2 = np$ if the semiconductor is in thermal equilibrium.

The scaled complete coupled system is constructed as follows. (We neglect the tilde-sign over the scaled variables.) The current through the diodes must be considered in Kirchhoff's current law. Consequently, the term $A_S j_S$ is added to Eq. (1.1), e.g.

$$A_C \frac{d}{dt} q_C(A_C^\top e, t) + A_R g(A_R^\top e, t) + A_L j_L + A_V j_V + A_S j_S = -A_I i_s(t), \tag{1.4}$$

$$\frac{d}{dt}\phi_L(j_L, t) - A_L^\top e = 0, \tag{1.5}$$

$$A_V^\top e = v_s(t). \tag{1.6}$$

In particular the matrix A_S denotes the semiconductor incidence matrix. Here,

$$j_{S,k} = \int_{\Gamma_{O,k}} (J_n + J_p - \varepsilon \partial_t \nabla\psi) \cdot v \, d\sigma. \tag{1.7}$$

I.e. the current is the integral over the current density $J_n + J_p$ plus the displacement current in normal direction v. Furthermore, the potentials of nodes which are connected to a semiconductor interface are introduced in the boundary conditions of the drift-diffusion equations (see also Fig. 1.2):

$$\psi(t,x) = \psi_{bi}(x) + (A_S^\top e(t))_k = U_T \log\left(\frac{\sqrt{C(x)^2 + 4n_i^2} + C(x)}{2n_i}\right) + (A_S^\top e(t))_k, \tag{1.8}$$

$$n(t,x) = \frac{1}{2}\left(\sqrt{C(x)^2 + 4n_i^2} + C(x)\right), \tag{1.9}$$

$$p(t,x) = \frac{1}{2}\left(\sqrt{C(x)^2 + 4n_i^2} - C(x)\right), \tag{1.10}$$

Fig. 1.1 Basic test circuit
with one diode

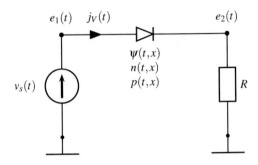

for $(t, x) \in [0, T] \times \Gamma_{O,k}$. Here, $\psi_{bi}(x)$ denotes the build-in potential and n_i the constant intrinsic concentration. All other parts of the boundary are isolation boundaries $\Gamma_I := \Gamma \setminus \Gamma_O$, where $\nabla \psi \cdot \nu = 0$, $J_n \cdot \nu = 0$ and $J_p \cdot \nu = 0$ holds. For a basic example consider the network in Fig. 1.1 where the network is described by

$$A_V = \begin{pmatrix} 1, 0 \end{pmatrix}^\top, \quad A_S = \begin{pmatrix} -1, 1 \end{pmatrix}^\top, \quad A_R = \begin{pmatrix} 0, 1 \end{pmatrix}^\top, \quad \text{and} \quad g(A_R^\top e, t) = \frac{1}{R} e_2(t).$$

The complete model forms a partial differential-algebraic equation (PDAE). The analytical and numerical analysis of such systems is subject to current research, see [7, 17, 46, 52]. The simulation of the complete coupled system is expensive and numerically difficult due to bad scaling of the drift-diffusion equations. The numerical issues can be significantly reduced by the unit scaling procedure discussed in [43]. That means we substitute

$$x = L\tilde{x}, \quad \psi = U_T \tilde{\psi}, \quad n = \|C\|_\infty \tilde{n}, \quad p = \|C\|_\infty \tilde{p}, \quad C = \|C\|_\infty \tilde{C},$$

$$J_n = \frac{q U_T \|C\|_\infty}{L} \mu_n \tilde{J}_n, \quad J_p = \frac{q U_T \|C\|_\infty}{L} \mu_p \tilde{J}_p, \quad n_i = \tilde{n}_i \|C\|_\infty,$$

where L denotes a specific length of the semiconductor (Fig. 1.2). The scaled drift-diffusion equations then read

$$\lambda \Delta \psi = n - p - C, \tag{1.11}$$

$$-\partial_t n + \nu_n \, \text{div} \, J_n = R(n, p), \tag{1.12}$$

$$\partial_t p + \nu_p \, \text{div} \, J_p = -R(n, p), \tag{1.13}$$

$$J_n = \nabla n - n \nabla \psi, \tag{1.14}$$

$$J_p = -\nabla p - p \nabla \psi, \tag{1.15}$$

where we omit the tilde for the scaled variables. The constants are given by $\lambda := \frac{\varepsilon U_T}{L^2 q \|C\|_\infty}$, $\nu_n := \frac{U_T \mu_n}{L^2}$ and $\nu_p := \frac{U_T \mu_p}{L^2}$, see e.g. [43].

Fig. 1.2 Sketch of a coupled system with one semiconductor. Here $\psi(t, x) = e_i(t) + \psi_{bi}(x)$, for all $(t, x) \in [0, T] \times \Gamma_{O,1}$

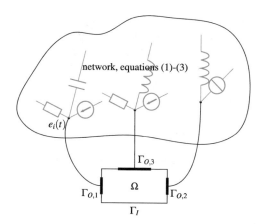

1.3   Simulation of the Full System

Classical approaches for the simulation of drift-diffusion equations (e.g. Gummel iterations [21]) approximate J_n and J_p by piecewise constant functions and then solve Eqs. (1.12) and (1.13) with respect to n and p explicitly. This helps reducing the computational effort and increases the numerical stability. For the model order reduction approach proposed in the present work this method has the disadvantage of introducing additional non-linearities, arising from the exponential structure of the Slotboom variables, see [46]. Subsequently we propose two finite element discretizations for the drift-diffusion system which with regard to coping with nonlinearities are advantageous from the MOR reduction point of view, and which together with the equations for the electrical network finally lead to large-scale nonlinear DAE model for the fully coupled system.

1.3.1   Standard Galerkin Finite Element Approach

Let $\mathscr{T}$ denote a regular triangulation of the domain Ω with gridwidth h, whose simplexes are denoted by T. In the classical Galerkin finite element method the functions ψ, n and p are approximated by piecewise linear and globally continuous functions, while J_n and J_p are approximated by patchwise-piecewise constant functions, e.g.

$$\psi(t, x) := \sum_{i=1}^{N} \psi_i(t)\phi_i(x), \quad n(t, x) := \sum_{i=1}^{N} n_i(t)\phi_i(x), \quad p(t, x) := \sum_{i=1}^{N} p_i(t)\phi_i(x),$$

$$J_n(t, x) := \sum_{i=1}^{N} J_{n,i}(t)\varphi_i(x), \quad J_p(t, x) := \sum_{i=1}^{N} J_{p,i}(t)\varphi_i(x),$$

where the functions $\{\phi_i\}$ and $\{\varphi_i\}$ are the corresponding ansatz functions, and N denotes the number of degrees of freedom. For ψ, n and p the coefficients corresponding to the boundary elements are prescribed using the Dirichlet boundary conditions. Note that the time is not discretized at this point which refers to the so-called method of lines. The finite element method leads to the following DAE for the unknown vector-valued functions of time ψ, n, p, J_n, J_p for each semiconductor:

$$0 = \lambda S\psi(t) + Mn(t) - Mp(t) - C_h + b_\psi(A_S^T e(t)),$$

$$-M\dot{n}(t) = -v_n D^\top J_n(t) + hR(n(t), p(t)),$$

$$M\dot{p}(t) = -v_p D^\top J_p(t) - hR(n(t), p(t)), \qquad (1.16)$$

$$0 = hJ_n(t) + Dn(t) - \text{diag}\left(Bn(t) + \tilde{b}_n\right) D\psi(t) + b_n,$$

$$0 = hJ_p(t) - Dp(t) - \text{diag}\left(Bp(t) + \tilde{b}_p\right) D\psi(t) + b_p,$$

where S, M and D, B are assembled finite element matrices. The matrix $\text{diag}(v)$ is diagonal with vector v forming the diagonal. The vectors $b_\psi(A_S^T e(t))$, b_n, $\tilde{b}_n$, b_p and $\tilde{b}_p$ implement the boundary conditions imposed on ψ, n and p through (1.8)–(1.10).

Discretization of the coupling condition for the current (1.7) completes the discretized system. In one spatial dimension we use

$$j_{S,k}(t) = \frac{aqU_T\|C\|_\infty}{L}\left(\mu_n J_{n,N}(t) + \mu_p J_{p,N}(t)\right) - \frac{a\varepsilon U_T}{Lh}\left(\dot{\psi}_N(t) - \dot{\psi}_{N-1}(t)\right),$$

1.3.2   Mixed Finite Element Approach

Since the electrical field represented by the (negative) gradient of the electrical potential ψ plays a dominant role in (1.11)–(1.15) and is present also in the coupling condition (1.7), we provide for it the additional variable $g_\psi = \nabla\psi$ leading to the following mixed formulation of the DD equations:

$$\lambda \, \text{div} \, g_\psi = n - p - C, \qquad (1.17)$$

$$-\partial_t n + v_n \, \text{div} \, J_n = R(n, p), \qquad (1.18)$$

$$\partial_t p + v_p \, \text{div} \, J_p = -R(n, p), \qquad (1.19)$$

$$g_\psi = \nabla\psi, \qquad (1.20)$$

$$J_n = \nabla n - n g_\psi, \qquad (1.21)$$

$$J_p = -\nabla p - p g_\psi. \qquad (1.22)$$

The weak formulation of (1.17)–(1.22) then reads: Find $\psi, n, p \in [0, T] \times L^2(\Omega)$ and $g_\psi, J_n, J_p \in [0, T] \times H_{0,N}(\text{div}, \Omega)$ such that

$$\lambda \int_\Omega \text{div}\, g_\psi \, \varphi = \int_\Omega (n - p)\, \varphi - \int_\Omega C\, \varphi, \tag{1.23}$$

$$-\int_\Omega \partial_t n \, \varphi + v_n \int_\Omega \text{div}\, J_n \, \varphi = \int_\Omega R(n, p)\, \varphi, \tag{1.24}$$

$$\int_\Omega \partial_t p \, \varphi + v_p \int_\Omega \text{div}\, J_p \, \varphi = -\int_\Omega R(n, p)\, \varphi, \tag{1.25}$$

$$\int_\Omega g_\psi \cdot \phi = -\int_\Omega \psi \, \text{div}\, \phi + \int_\Gamma \psi \, \phi \cdot v, \tag{1.26}$$

$$\int_\Omega J_n \cdot \phi = -\int_\Omega n \, \text{div}\, \phi + \int_\Gamma n \, \phi \cdot v - \int_\Omega n \, g_\psi \cdot \phi, \tag{1.27}$$

$$\int_\Omega J_p \cdot \phi = \int_\Omega p \, \text{div}\, \phi - \int_\Gamma p \, \phi \cdot v - \int_\Omega p \, g_\psi \cdot \phi, \tag{1.28}$$

are satisfied for all $\varphi \in L^2(\Omega)$ and $\phi \in H_{0,N}(\text{div}, \Omega)$ where the space $H_{0,N}(\text{div}, \Omega)$ is defined by

$$H(\text{div}, \Omega) := \{ y \in L^2(\Omega)^d : \text{div}\, y \in L^2(\Omega) \},$$

$$H_{0,N}(\text{div}, \Omega) := \{ y \in H(\text{div}, \Omega) : y \cdot v = 0 \text{ on } \Gamma_I \}.$$

Consequently, the boundary integrals on the right hand sides in Eqs. (1.26)–(1.28) reduce to integrals over the interfaces $\Gamma_{0,k}$, where the values of ψ, n and p are determined by the Dirichlet boundary conditions (1.8)–(1.10). We note that, in contrast to the standard weak form associated with (1.11)–(1.15), the Dirichlet boundary values are naturally included in the weak formulation (1.23)–(1.28) and the Neumann boundary conditions have to be included in the space definitions. This is advantageous in the context of POD model order reduction since the non-homogeneous boundary conditions (1.8)–(1.10) are not present in the space definitions.

Here, Eqs. (1.23)–(1.28) are discretized in space with Raviart-Thomas finite elements of degree 0 (RT_0), alternative discretization schemes for the mixed problem are presented in [8]. To describe the RT_0-approach for $d = 2$ spatial dimensions, let $\mathcal{T}$ be a triangulation of Ω and let $\mathcal{E}$ be the set of all edges. Let $\mathcal{E}_I := \{ E \in \mathcal{E} : E \subset \bar{\Gamma}_I \}$ be the set of edges at the isolation (Neumann) boundaries. The potential and the concentrations are approximated in space by piecewise constant functions

$$\psi^h(t), n^h(t), p^h(t) \in L_h := \{ y \in L^2(\Omega) : y|_T(x) = c_T, \forall T \in \mathcal{T} \},$$

with ansatz functions $\{\varphi_i\}_{i=1,\ldots,N}$ and the discrete fluxes $g_\psi^h(t)$, $J_n^h(t)$ and $J_p^h(t)$ are elements of the space

$$RT_0 := \{y : \Omega \to \mathbb{R}^d : y|_T(x) = a_T + b_T x, \ a_T \in \mathbb{R}^d, \ b_T \in \mathbb{R}, \ [y]_E \cdot \nu_E = 0,$$

$$\text{for all inner edges } E\}.$$

Here, $[y]_E$ denotes the jump $y|_{T_+} - y|_{T_-}$ over a shared edge E of the elements T_+ and T_-. The continuity assumption yields $RT_0 \subset H(\mathrm{div}, \Omega)$. We set

$$H_{h,0,N}(\mathrm{div}, \Omega) := (RT_0 \cap H_{0,N}(\mathrm{div}, \Omega)) \subset H_{0,N}(\mathrm{div}, \Omega).$$

Then it can be shown, that $H_{h,0,N}$ posses an edge-oriented basis $\{\phi_j\}_{j=1,\ldots,M}$. We use the following finite element ansatz in (1.23)–(1.28)

$$
\left.
\begin{aligned}
\psi^h(t,x) &= \sum_{i=1}^{N} \psi_i(t)\varphi_i(x), & g_\psi^h(t,x) &= \sum_{j=1}^{M} g_{\psi,j}(t)\phi_j(x), \\
n^h(t,x) &= \sum_{i=1}^{N} n_i(t)\varphi_i(x), & J_n^h(t,x) &= \sum_{j=1}^{M} J_{n,j}(t)\phi_j(x), \\
p^h(t,x) &= \sum_{i=1}^{N} p_i(t)\varphi_i(x), & J_p^h(t,x) &= \sum_{j=1}^{M} J_{p,j}(t)\phi_j(x),
\end{aligned}
\right\} \tag{1.29}
$$

where $N := |\mathcal{T}|$, i.e. the number of elements of $\mathcal{T}$, and $M := |\mathcal{E}| - |\mathcal{E}_N|$, i.e. the number of inner and Dirichlet boundary edges.

This in (1.23)–(1.28) yields

$$\lambda \sum_{j=1}^{M} g_{\psi,j}(t) \int_\Omega \mathrm{div}\, \phi_j\, \varphi_k - \sum_{i=1}^{N} (n_i(t) - p_i(t)) \int_\Omega \varphi_i\, \varphi_k = -\int_\Omega C\, \varphi_k,$$

$$-\sum_{i=1}^{N} \dot{n}_i(t) \int_\Omega \varphi_i\, \varphi_k + \nu_n \sum_{j=1}^{M} J_{n,j}(t) \int_\Omega \mathrm{div}\, \phi_j\, \varphi_k - \int_\Omega R(n^h, p^h)\, \varphi_k = 0,$$

$$\sum_{i=1}^{N} \dot{p}_i(t) \int_\Omega \varphi_i\, \varphi_k + \nu_p \sum_{j=1}^{M} J_{p,j}(t) \int_\Omega \mathrm{div}\, \phi_j\, \varphi_k + \int_\Omega R(n^h, p^h)\, \varphi_k = 0,$$

$$\sum_{j=1}^{M} g_{\psi,j}(t) \int_\Omega \phi_j \cdot \phi_l + \sum_{i=1}^{N} \psi_i(t) \int_\Omega \varphi_i\, \mathrm{div}\, \phi_l = \int_\Gamma \psi^h\, \phi_l \cdot \nu,$$

$$\sum_{j=1}^{M} J_{n,j}(t) \int_{\Omega} \phi_j \cdot \phi_l + \sum_{i=1}^{N} n_i(t) \int_{\Omega} \varphi_i \text{ div } \phi_l + \int_{\Omega} n^h g_\psi^h \cdot \phi_l = \int_{\Gamma} n^h \phi_l \cdot \nu,$$

$$\sum_{j=1}^{M} J_{p,j}(t) \int_{\Omega} \phi_j \cdot \phi_l - \sum_{i=1}^{N} p_i(t) \int_{\Omega} \varphi_i \text{ div } \phi_l + \int_{\Omega} p^h g_\psi^h \cdot \phi_l = - \int_{\Gamma} p^h \phi_l \cdot \nu,$$

which represents a nonlinear, large and sparse DAE for the approximation of the functions $\psi, n, p, g_\psi, J_n,$ and J_p. In matrix notation it reads

$$\begin{pmatrix} 0 \\ -M_L \dot{n}(t) \\ M_L \dot{p}(t) \\ 0 \\ 0 \\ 0 \end{pmatrix} + \underbrace{\begin{pmatrix} -M_L & M_L & \lambda D & & & \\ & & & \nu_n D & & \\ & & & & \nu_p D & \\ D^\top & & & M_H & & \\ & D^\top & & & M_H & \\ & & -D^\top & & & M_H \end{pmatrix}}_{A_{FEM}} \begin{pmatrix} \psi(t) \\ n(t) \\ p(t) \\ g_\psi(t) \\ J_n(t) \\ J_p(t) \end{pmatrix}$$

$$+ \mathscr{F}(n^h, p^h, g_\psi^h) = b(A_S^T e(t)),$$

with

$$\mathscr{F}(n^h, p^h, g_\psi^h) := \begin{pmatrix} 0 \\ -\int_\Omega R(n^h, p^h) \varphi \\ \int_\Omega R(n^h, p^h) \varphi \\ 0 \\ \int_\Omega n^h g_\psi^h \cdot \phi \\ \int_\Omega p^h g_\psi^h \cdot \phi \end{pmatrix}, \quad b := \begin{pmatrix} -\int_\Omega C \varphi \\ 0 \\ 0 \\ \int_\Gamma \psi^h (A_S^T e(t)) \phi \cdot \nu \\ \int_\Gamma n^h \phi \cdot \nu \\ -\int_\Gamma p^h \phi \cdot \nu \end{pmatrix},$$

$$\tag{1.30}$$

and

$$\int_\Omega R(n^h, p^h) \varphi := \begin{pmatrix} \int_\Omega R(n^h, p^h) \varphi_1 \\ \vdots \\ \int_\Omega R(n^h, p^h) \varphi_N \end{pmatrix}.$$

All other integrals in $\mathscr{F}$ and b are defined analogously. The matrices $M_L \in \mathbb{R}^{N \times N}$ and $M_H \in \mathbb{R}^{M \times M}$ are mass matrices in the spaces L_h and $H_{h,0,N}$, respectively, and $D \in \mathbb{R}^{N \times M}$. The final DAE for the mixed finite element discretization now takes the form

Problem 1.3.1 (Full Model)

$$A_C \frac{d}{dt} q_C(A_C^\top e(t), t) + A_R g(A_R^\top e(t), t) + A_L j_L(t) + A_V j_V(t)$$

$$+ A_S j_S(t) + A_I i_s(t) = 0, \qquad (1.31)$$

$$\frac{d}{dt} \phi_L(j_L(t), t) - A_L^\top e(t) = 0, \qquad (1.32)$$

$$A_V^\top e(t) - v_s(t) = 0, \qquad (1.33)$$

$$j_S(t) - C_1 J_n(t) - C_2 J_p(t) - C_3 \dot{g}_\psi(t) = 0, \qquad (1.34)$$

$$\begin{pmatrix} 0 \\ -M_L \dot{n}(t) \\ M_L \dot{p}(t) \\ 0 \\ 0 \\ 0 \end{pmatrix} + A_{FEM} \begin{pmatrix} \psi(t) \\ n(t) \\ p(t) \\ g_\psi(t) \\ J_n(t) \\ J_p(t) \end{pmatrix} + \mathscr{F}(n^h, p^h, g_\psi^h) - b(A_S^T e(t)) = 0, \qquad (1.35)$$

where (1.34) represents the discretized linear coupling condition (1.7).

We present numerical computations for the basic test circuit with one diode depicted in Fig. 1.1, where the model parameters are presented in Table 1.1. The input $v_s(t)$ is chosen to be sinusoidal with amplitude 5 V. The numerical results in Fig. 1.3 show the capacitive effect of the diode for high input frequencies. Similar results are obtained in [44] using the simulator MECS.

The discretized equations are implemented in MATLAB, and the DASPK software package [34] is used to integrate the high-dimensional DAE. Initial values are stationary states obtained by setting all time derivatives to 0. In order to solve the Newton systems which arise from the BDF method efficiently, one may reorder the variables of the sparse system with respect to minimal bandwidth. Then, one can use the internal DASPK routines for the solution of the linear systems. Alternatively one can implement the preconditioning subroutine of DASPK using a direct sparse solver. Note that for both strategies we only need to calculate the reordering matrices once, since the sparsity structure remains constant.

Table 1.1 Diode model parameters

Parameter	Value	Parameter	Value
L	10^{-4} cm	ε	$1.03545 \cdot 10^{-12}$ F/cm
U_T	0.0259 V	n_i	$1.4 \cdot 10^{10}$ 1/cm^3
μ_n	1350 cm^2/(V s)	τ_n	$330 \cdot 10^{-9}$ s
μ_p	480 cm^2/(V s)	τ_p	$33 \cdot 10^{-9}$ s
a	10^{-5} cm^2	$C(x),\ x < L/2$	$-9.94 \cdot 10^{15}$ 1/cm^3
		$C(x),\ x \geq L/2$	$4.06 \cdot 10^{18}$ 1/cm^3

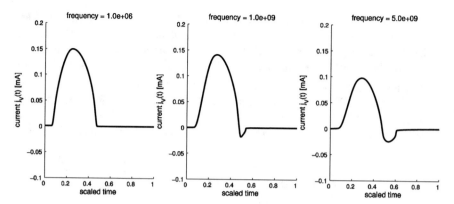

Fig. 1.3 Current j_V through the basic network for input frequencies 1 MHz, 1 GHz and 5 GHz. The capacitive effect is clearly demonstrated

1.4   Model Order Reduction Using POD

We now use proper orthogonal decomposition (POD) to construct low-dimensional surrogate models for the drift-diffusion equations. The idea consists in replacing the large number of local model-independent ansatz and test functions $\{\phi_i\}$, $\{\varphi_j\}$ in the finite element approximation of the drift-diffusion systems by only a few nonlocal model-dependent ansatz functions for the respective variables.

The snapshot variant of POD introduced in [45] works as follows. We run a simulation of the unreduced system and collect l snapshots $\psi^h(t_k, \cdot)$, $n^h(t_k, \cdot)$, $p^h(t_k, \cdot)$, $g_\psi^h(t_k, \cdot)$, $J_n^h(t_k, \cdot)$, $J_p^h(t_k, \cdot)$ at time instances $t_k \in \{t_1, \ldots, t_l\} \subset [0, T]$. The optimal selection of the time instances is not considered here. We use the time instances delivered by the DAE integrator.

Since every component of the state vector $y := (\psi, n, p, g_\psi, J_n, J_p)$ has its own physical meaning we apply POD MOR to each component separately. Among other things this approach has the advantage of yielding a block-dense model and the approximation quality of each component is adapted individually.

Let X denote a Hilbert space and let $y^h : [0, T] \times X \to \mathbb{R}^r$ with some $r \in \mathbb{N}$. The Galerkin formulation (1.29) yields $y^h(t, \cdot) \in X_h := \text{span}\{\phi_1^X, \ldots, \phi_n^X\}$, where $\{\phi_j^X\}_{1 \le j \le n}$ denote n linearly independent elements of X. The idea of POD consists in finding a basis $\{u^1, \ldots, u^m\}$ of the span of the snapshots

$$\text{span}\left\{ y^h(t_k, \cdot) = \sum_{i=1}^n y_i^{h,k} \phi_i^X(\cdot), \text{ with } k = 1, \ldots, l \right\}$$

satisfying

$$\{u^1, \ldots, u^s\} = \underset{\{v^1, \ldots, v^s\} \subset X}{\arg\min} \sum_{k=1}^l \left\| y^h(t_k, \cdot) - \sum_{i=1}^s \langle y^h(t_k, \cdot), v^i(\cdot) \rangle_X v^i(\cdot) \right\|_X^2,$$

for $1 \leq s \leq m$, where $1 \leq m \leq l$. The functions $\{u^i\}_{1 \leq i \leq s}$ are orthonormal in X and can be obtained with the help of the singular value decomposition (SVD) as follows.

Let the matrix $Y := (y^{h,1}, \ldots, y^{h,l}) \in \mathbb{R}^{n \times l}$ contain as columns the coefficient vectors of the snapshots. Furthermore, let $M := (\langle \phi_i^X, \phi_j^X \rangle_X)_{1 \leq i,j \leq n}$ be the positive definite mass matrix with its Cholesky factorization $M = LL^\top$. Let $(\tilde{U}, \Sigma, \tilde{V})$ denote the SVD of $\tilde{Y} := L^\top Y$, i.e. $\tilde{Y} = \tilde{U} \Sigma \tilde{V}^\top$ with $\tilde{U} \in \mathbb{R}^{n \times n}$, $\tilde{V} \in \mathbb{R}^{l \times l}$, and a matrix $\Sigma \in \mathbb{R}^{n \times l}$ containing the singular values $\sigma_1 \geq \sigma_2 \geq \ldots \geq \sigma_m > \sigma_{m+1} = \ldots = \sigma_l = 0$. We set $U := L^{-\top} \tilde{U}_{(:, 1:s)}$. Then, the s-dimensional POD basis is given by

$$\text{span} \left\{ u^i(\cdot) = \sum_{j=1}^n U_{ji} \phi_j^X(\cdot), \ i = 1, \ldots, s \right\}.$$

The information content of $\{u^1, \ldots, u^s\}$ with respect to the scalar product $\langle \cdot, \cdot \rangle_X$ with

$$0 \leq \Delta(s) = \sqrt{\frac{\sum_{i=s+1}^m \sigma_i^2}{\sum_{i=1}^m \sigma_i^2}} \leq 1, \tag{1.36}$$

is given by $1 - \Delta(s)$. Here $\Delta(s)$ measures the lack of information of $\{u^1, \ldots, u^s\}$ with respect to $\text{span}\{y^h(t_1, \cdot), \ldots, y^h(t_l, \cdot)\}$. An extended introduction to POD can be found in [36].

The POD basis functions are now used as trial and test functions in the Galerkin method.

If the snapshots satisfy inhomogeneous Dirichlet boundary conditions, as in (1.16), POD is performed for

$$\tilde{\psi}(t) = \psi(t) - \psi_r(t), \quad \tilde{n}(t) = n(t) - n_r(t), \quad \tilde{p}(t) = p(t) - p_r(t),$$

with ψ_r, n_r, p_r denoting reference functions satisfying the Dirichlet boundary conditions required for ψ, n and p. This guarantees that the POD basis admits homogeneous boundary conditions on the Dirichlet boundary.

In the case of the mixed finite element approach the introduction of a reference state is not necessary, since the boundary values are included more naturally through the variational formulation. The time-snapshot POD procedure then delivers Galerkin ansatz spaces for ψ, n, p, g_ψ, J_n and J_p. This leads to the ansatz

$$\left. \begin{array}{ll} \psi^{POD}(t) = U_\psi \gamma_\psi(t), & g_\psi^{POD}(t) = U_{g_\psi} \gamma_{g_\psi}(t), \\[2mm] n^{POD}(t) = U_n \gamma_n(t), & J_n^{POD}(t) = U_{J_n} \gamma_{J_n}(t), \\[2mm] p^{POD}(t) = U_p \gamma_p(t), & J_p^{POD}(t) = U_{J_p} \gamma_{J_p}(t). \end{array} \right\} \tag{1.37}$$

The injection matrices

$$U_\psi \in \mathbb{R}^{N \times s_\psi}, \qquad\qquad U_n \in \mathbb{R}^{N \times s_n}, \qquad\qquad U_p \in \mathbb{R}^{N \times s_p},$$

$$U_{g_\psi} \in \mathbb{R}^{M \times s_{g_\psi}}, \qquad\qquad U_{J_n} \in \mathbb{R}^{M \times s_{J_n}}, \qquad\qquad U_{J_p} \in \mathbb{R}^{M \times s_{J_p}},$$

contain the (time independent) POD basis functions, the vectors $\gamma_{(\cdot)}$ the corresponding time-variant coefficients. The numbers $s_{(\cdot)}$ are the respective number of POD basis functions included. Assembling the POD system yields the DAE

$$\begin{pmatrix} 0 \\ -\dot\gamma_n(t) \\ \dot\gamma_p(t) \\ 0 \\ 0 \\ 0 \end{pmatrix} + A_{POD} \begin{pmatrix} \gamma_\psi(t) \\ \gamma_n(t) \\ \gamma_p(t) \\ \gamma_{g_\psi}(t) \\ \gamma_{J_n}(t) \\ \gamma_{J_p}(t) \end{pmatrix} + U^\top \mathscr{F}(n^{POD}, p^{POD}, g_\psi^{POD}) = U^\top b(A_S^T e(t)),$$

with

$$A_{POD} = U^\top A_{FEM} U$$

$$= \begin{pmatrix} & -U_\psi^\top M_L U_n & U_\psi^\top M_L U_p & \lambda U_\psi^\top D U_{g_\psi} & & \\ & & & & v_n U_n^\top D U_{J_n} & \\ & & & & & v_p U_p^\top D U_{J_p} \\ U_{g_\psi}^\top D^\top U_\psi & & & I & & \\ & U_{J_n}^\top D^\top U_n & & & I & \\ & & -U_{J_p}^\top D^\top U_p & & & I \end{pmatrix}$$

and $U = \operatorname{diag}(U_\psi, U_n, U_p, U_{g_\psi}, U_{J_n}, U_{J_p})$. Note that we exploit the orthogonality of the POD basis functions, e.g. $U_n^\top M_L U_n = U_p^\top M_L U_p = I_{N \times N}$ and $U_{g_\psi}^\top M_H U_{g_\psi} = U_{J_n}^\top M_H U_{J_n} = U_{J_p}^\top M_H U_{J_p} = I_{M \times M}$. The arguments of the nonlinear functional have to be interpreted as functions in space.

All matrix-matrix multiplications are calculated in an offline phase. The nonlinear functional $\mathscr{F}$ has to be evaluated online. The reduced model for the network now reads

Problem 1.4.1 (POD MOR Surrogate)

$$A_C \frac{d}{dt} q_C(A_C^\top e(t), t) + A_R g(A_R^\top e(t), t) + A_L j_L(t) + A_V j_V(t)$$

$$+ A_S j_S(t) + A_I i_S(t) = 0, \tag{1.38}$$

$$\frac{d}{dt}\phi_L(j_L(t),t) - A_L^\top e(t) = 0,$$

$$(1.39)$$

$$A_V^\top e(t) - v_s(t) = 0,$$

$$(1.40)$$

$$j_S(t) - C_1 U_{J_n}\gamma_{J_n}(t) - C_2 U_{J_p}\gamma_{J_p}(t) - C_3 U_{g_\psi}\dot{\gamma}_{g_\psi}(t) = 0,$$

$$(1.41)$$

$$\begin{pmatrix} 0 \\ -\dot{\gamma}_n(t) \\ \dot{\gamma}_p(t) \\ 0 \\ 0 \\ 0 \end{pmatrix} + A_{POD} \begin{pmatrix} \gamma_\psi(t) \\ \gamma_n(t) \\ \gamma_p(t) \\ \gamma_{g_\psi}(t) \\ \gamma_{J_n}(t) \\ \gamma_{J_p}(t) \end{pmatrix} + U^\top \mathscr{F}(n^{POD}, p^{POD}, g_\psi^{POD}) - U^\top b(A_S^\top e(t)) = 0.$$

$$(1.42)$$

1.4.1 Numerical Investigation

We now present numerical examples for POD MOR of the basic test circuit in Fig. 1.1 and validate the reduced model at a fixed reference frequency of 10^{10} Hz. Figure 1.4 (left) shows the development of the error between the reduced and the unreduced numerical solutions, plotted over the neglected information Δ, see (1.36), which is measured by the relative error between the non-reduced states ψ, n, p, J_n, J_p and their projections onto the respective reduced state space. The number of POD basis functions for each variable is chosen such that the indicated approximation quality is reached, i.e. $\Delta := \Delta_\psi \simeq \Delta_n \simeq \Delta_p \simeq \Delta_{g_\psi} \simeq \Delta_{J_n} \simeq \Delta_{J_p}$. Since we compute all POD basis functions anyway, this procedure does not involve any additional costs.

In Fig. 1.4 (right) the simulation times are plotted versus the neglected information Δ. As one also can see, the simulation based on standard finite elements takes twice as long as if based on RT elements. However, this difference is not observed for the simulation of the corresponding reduced models.

Figure 1.5 shows the total number of singular vectors $k = k_\psi + k_n + k_p + k_{J_n} + k_{J_p}$ required in the POD model to guarantee a given state space cut-off error Δ. While the number of singular vectors included increases only linearly, the cut-off error tends to zero exponentially.

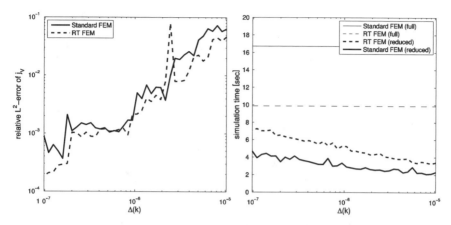

Fig. 1.4 *Left*: L_2 error of j_V between reduced and unreduced problem, both for standard and Raviart-Thomas FEM. *Right*: Time consumption for simulation runs for *left figure*. The *fine lines* indicate the time consumption for the simulation of the original full system

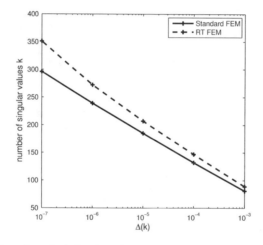

Fig. 1.5 The number of required singular values grows only logarithmically with the requested accuracy

1.4.2 Numerical Investigation, Position of the Semiconductor in the Network

Finally we note that the presented reduction method accounts for the position of the semiconductors in a given network in that it provides reduced-order models which for identical semiconductors may be different depending on the location of the semiconductors in the network. The POD basis functions of two identical semiconductors may be different due to their different operating states. To demonstrate this fact, we consider the rectifier network in Fig. 1.6 (left). Simulation results are

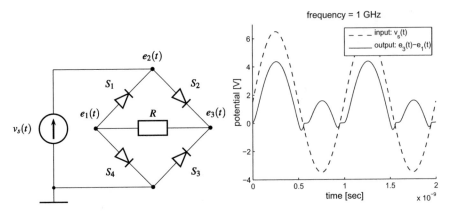

Fig. 1.6 *Left*: Rectifier network. *Right*: Simulation results for the rectifier network. The input v_s is sinusoidal with frequency 1 GHz and offset $+1.5$ V

Table 1.2 Distances
between reduced models in
the rectifier network

Δ	$d(U^1, U^2)$	$d(U^1, U^3)$
10^{-4}	0.61288	$5.373 \cdot 10^{-8}$
10^{-5}	0.50766	$4.712 \cdot 10^{-8}$
10^{-6}	0.45492	$2.767 \cdot 10^{-7}$
10^{-7}	0.54834	$1.211 \cdot 10^{-6}$

plotted in Fig. 1.6 (right). The distance between the spaces U^1 and U^2 which are spanned, e.g., by the POD-functions U^1_ψ of the diode S_1 and U^2_ψ of the diode S_2 respectively, is measured by

$$d(U^1, U^2) := \max_{\substack{u \in U^1 \\ \|u\|_2 = 1}} \min_{\substack{v \in U^2 \\ \|v\|_2 = 1}} \|u - v\|_2.$$

Exploiting the orthonormality of the bases U^1_ψ and U^2_ψ and using a Lagrange framework, we find

$$d(U^1, U^2) = \sqrt{2 - 2\sqrt{\lambda}},$$

where λ is the smallest eigenvalue of the positive definite matrix $SS^\top$ with $S_{ij} = \langle u^1_{\psi,i}, u^2_{\psi,j} \rangle_2$. The distances for the rectifier network are given in Table 1.2. While the reduced model for the diodes S_1 and S_3 are almost equal, the models for the diodes S_1 and S_2 are significantly different. Similar results are obtained for the reduction of n, p, etc.

1.4.3   MOR for the Nonlinearity with DEIM

The nonlinear function $\mathscr{F}$ in (1.42) has to be evaluated online which means that the computational complexity of the reduced-order model still depends on the number of unknowns of the unreduced model. A reduction method for the nonlinearity is given by Discrete Empirical Interpolation (DEIM) [10]. This method is motivated by the following observation. The nonlinearity in (1.42), see also (1.30), is given by

$$
U^{\top}\mathscr{F}(U\gamma(t)) = \begin{pmatrix}
0 \\
U_n^{\top} F_n(U_n\gamma_n(t), U_p\gamma_p(t)) \\
U_p^{\top} F_p(U_n\gamma_n(t), U_p\gamma_p(t)) \\
0 \\
U_{J_n}^{\top} F_{J_n}(U_n\gamma_n(t), U_{g_\psi}\gamma_{g_\psi}(t)) \\
U_{J_p}^{\top} F_{J_p}(U_n\gamma_p(t), U_{g_\psi}\gamma_{g_\psi}(t))
\end{pmatrix},
$$

see e.g. [23]. The subsequent considerations apply for each block component of $\mathscr{F}$. For the sake of presentation we only consider the second block

$$
\underbrace{U_n^{\top}}_{\text{size } s_n\times\mathbf{N}} \quad \underbrace{F_n}_{\mathbf{N}\text{ evaluations}} \quad (\underbrace{U_n}_{\text{size } \mathbf{N}\times s_n} \gamma_n(t), \underbrace{U_p}_{\text{size } \mathbf{N}\times s_p} \gamma_p(t)), \tag{1.43}
$$

and its derivative with respect to γ_p,

$$
\underbrace{U_n^{\top}}_{\text{size } s_n\times\mathbf{N}} \quad \underbrace{\frac{\partial F_n}{\partial p}(U_n\gamma_n(t), U_p\gamma_p(t))}_{\text{size } \mathbf{N}\times\mathbf{N},\text{ sparse}} \underbrace{U_p}_{\text{size } \mathbf{N}\times s_p} .
$$

Here, the matrices $U_{(\cdot)}$ are dense and the Jacobian of F_n is sparse. The evaluation of (1.43) is of computational complexity $O(N)$. Furthermore, we need to multiply large dense matrices in the evaluation of the Jacobian. Thus, the POD model order reduction may become inefficient.

   To overcome this problem, we apply Discrete Empirical Interpolation Method (DEIM) proposed in [10], which we now describe briefly. The snapshots $\psi^h(t_k, \cdot)$, $n^h(t_k, \cdot)$, $p^h(t_k, \cdot)$, $g_\psi^h(t_k, \cdot)$, $J_n^h(t_k, \cdot)$, $J_p^h(t_k, \cdot)$ are collected at time instances $t_k \in \{t_1, \ldots, t_l\} \subset [0, T]$ as before. Additionally, we collect snapshots $\{F_n(n(t_k), p(t_k))\}$ of the nonlinearity. DEIM approximates the projected function (1.43) such that

$$
U_n^{\top} F_n(U_n\gamma_n(t), U_p\gamma_p(t)) \approx (U_n^{\top} V_n(P_n^{\top} V_n)^{-1})P_n^{\top} F_n(U_n\gamma_n(t), U_p\gamma_p(t)),
$$

where $V_n \in \mathbb{R}^{N\times\tau_n}$ contains the first τ_n POD basis functions of the space spanned by the snapshots $\{F_n(n(t_k), p(t_k))\}$ associated with the largest singular values. The selection matrix $P_n = (e_{\rho_1}, \ldots, e_{\rho_{\tau_n}}) \in \mathbb{R}^{N\times\tau_n}$ selects the rows of F_n corresponding to the so-called DEIM indices $\rho_1, \ldots, \rho_{\tau_n}$ which are chosen such that the growth of a global error bound is limited and $P_n^\top V_n$ is regular, see [10] for details.

The matrix $W_n := (U_n^\top V_n (P_n^\top V_n)^{-1}) \in \mathbb{R}^{s_n \times \tau_n}$ as well as the whole interpolation method is calculated in an offline phase. In the simulation of the reduced-order model we instead of (1.43) evaluate:

$$\underbrace{W_n}_{\text{size } s_n \times \tau_n} \quad \underbrace{P_n^\top F_n}_{\tau_n \text{ evaluations}} \quad (\underbrace{U_n}_{\text{size } N \times s_n} \gamma_n(t), \underbrace{U_p}_{\text{size } N \times s_p} \gamma_p(t) \ ), \tag{1.44}$$

with derivative

$$\underbrace{W_n^\top}_{\text{size } s_n \times \tau_n} \quad \underbrace{\frac{\partial P_n^\top F_n}{\partial p}(U_n \gamma_n(t), U_p \gamma_p(t))}_{\text{size } \tau_n \times N, \text{ sparse}} \quad \underbrace{U_p}_{\text{size } N \times s_p} .$$

In the applied finite element method a single functional component of F_n only depends on a small constant number $c \in \mathbb{N}$ components of $U_n \gamma_n(t)$. Thus, the matrix-matrix multiplication in the derivative does not really depend on N since the number of entries per row in the Jacobian is at most c.

But there is still a dependence on N, namely the calculation of $U_n \gamma_n(t)$. To overcome this dependency we identify the required components of the vector $U_n \gamma_n(t)$ for the evaluation of $P_n^\top F_n$. This is done by defining selection matrices $Q_{n,n} \in \mathbb{R}^{c\tau_n \times s_n}$, $Q_{n,p} \in \mathbb{R}^{c\tau_p \times s_p}$ such that

$$P_n^\top F_n(U_n \gamma_n(t), U_p \gamma_p(t)) = \hat{F}_n(Q_{n,n} U_n \gamma_n(t), Q_{n,p} U_p \gamma_p(t)),$$

where $\hat{F}_n$ denotes the functional components of F_n selected by P_n restricted to the arguments selected by $Q_{n,n}$ and $Q_{n,p}$.

Supposed that $\tau_n \approx s_n \ll N$ we obtain a reduced-order model which does not depend on N any more.

1.4.4 Numerical Implementation and Results with DEIM

We again use the basic test circuit with a single 1-dimensional diode depicted in Fig. 1.1. The parameters of the diode are summarized in [23]. The input $v_s(t)$ is chosen to be sinusoidal with amplitude 5 V. In the sequel the frequency of the voltage source will be considered as a model parameter.

We first validate the reduced model at a fixed reference frequency of $5 \cdot 10^9$ Hz. Figure 1.7 shows the development of the relative error between the POD reduced, the POD-DEIM reduced and the unreduced numerical solutions, plotted over the lack of information Δ of the POD basis functions with respect to the space spanned by the snapshots. The figure shows that the approximation quality of the POD-DEIM

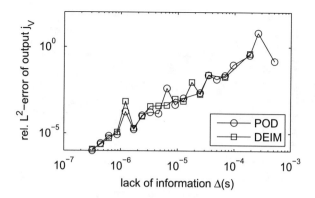

Fig. 1.7 Relative error between DEIM-reduced and unreduced nonlinearity at the fixed frequency $5 \cdot 10^9$ Hz

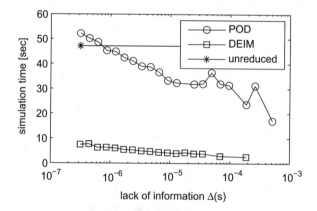

Fig. 1.8 Time consumption for simulation runs of Fig. 1.7. The *horizontal line* indicates the time consumption for the simulation of the original full system

reduced model is comparable with the more expensive POD reduced model. The number of POD basis functions $s_{(\cdot)}$ for each variable is chosen such that the indicated approximation quality is reached, i.e. $\Delta := \Delta_\psi \simeq \Delta_n \simeq \Delta_p \simeq \Delta_{g_\psi} \simeq \Delta_{J_n} \simeq \Delta_{J_p}$. The numbers $\tau_{(\cdot)}$ of POD-DEIM basis functions are chosen likewise.

In Fig. 1.8 the simulation times are plotted versus the lack of information Δ. The POD reduced-order model does not reduce the simulation times significantly for the chosen parameters. The reason for this is its dependency on the number of variables of the unreduced system. Here, the unreduced system contains 1000 finite elements which yields 12,012 unknowns. The POD-DEIM reduced-order model behaves very well and leads to a reduction in simulation time of about 90% without reducing the accuracy of the reduced model. However, we have to report a minor drawback; not all tested reduced models converge for large $\Delta(s) \geq 3 \cdot 10^{-5}$. This is indicated in the figures by missing squares. This effect is even more pronounced for

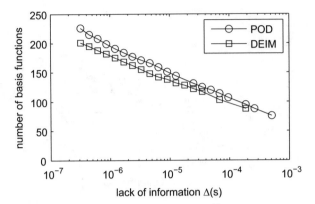

Fig. 1.9 The number of required POD basis function and DEIM interpolation indices grows only logarithmically with the requested information content

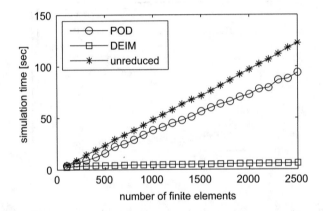

Fig. 1.10 Computation times of the unreduced and the reduced-order models plotted versus the number of finite elements

spatially two-dimensional semiconductors. It seems to be caused by the fact, that only a sufficiently large POD basis captured the physics of the semiconductors well enough.

In Fig. 1.9 we plot the corresponding total number of required POD basis functions. It can be seen that with the number of POD basis functions increasing linearly, the lack of information tends to zero exponentially. Furthermore, the number of DEIM interpolation indices behaves in the same way.

In Fig. 1.10 we investigate the dependence of the reduced models on the number of finite elements N. One sees that the simulation times of the unreduced model depends linearly on N. The POD reduced-order model still depends on N linearly with a smaller constant. The dependence on N of our POD-DEIM implementation is negligible.

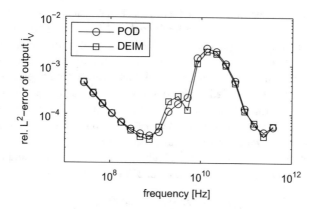

Fig. 1.11 The reduced models are compared with the unreduced model at various input frequencies

Finally, we analyze in Fig. 1.11 the behaviour of the models with respect to parameter changes. We consider the frequency of the sinusoidal input voltage as model parameter. The reduced-order models are created based on snapshots gathered in a full simulation at a frequency of $5 \cdot 10^9$ Hz. We see that the POD model and the POD-DEIM model behave very similarly. The adaptive enlargement of the POD basis using the residual greedy approach of [33] is discussed in the next section based on the results presented in [23].

Summarizing all numerical results we conclude that the significantly faster POD-DEIM reduction method yields a reduced-order model with the same qualitative behaviour as the reduced model obtained by classical POD MOR.

1.5 Residual-Based Sampling

Although POD model order reduction often works well, one has to keep in mind that the reduced system depends on the specific inputs and parameters used to generate the snapshots. A possible remedy consists in performing simulations over a certain input and/or parameter sample and then to collect all simulations in a global snapshot matrix $Y := [Y^1, Y^2, \ldots]$. Here, each Y^i represents the snapshots taken for a certain input resp. parameter.

In this section we propose a strategy to choose inputs/parameters in order to obtain a reduced model, which is valid over the whole input/parameter range. Possible parameters are physical constants of the semiconductors (e.g. length, permeability, doping) and parameters of the network elements (e.g. frequency of sinusoidal voltage sources, value of resistances). We do not distinguish between inputs and parameters of the model.

Let there be $r \in \mathbb{N}$ parameters and let the space of considered parameters be given as a bounded set $\mathscr{P} \subset \mathbb{R}^r$. We construct the reduced model based on snapshots from a simulation at a reference parameter $\omega_1 \in \mathscr{P}$. One expects that the reduced model approximates the unreduced model well in a small neighborhood of ω_1, but one cannot expect that the reduced model is valid over the complete parameter set $\mathscr{P}$. In order to create a suitable reduced-order model we consider additional snapshots which are obtained from simulations at parameters $\omega_2, \omega_3, \ldots \in \mathscr{P}$. The iterative selection of ω_{k+1} at a step k is called parameter sampling. Let P_k denote the set of selected reference parameters, $P_k := \{\omega_1, \omega_2, \ldots, \omega_k\} \subset \mathscr{P}$.

We neglect the discretization error of the finite element method and its influence on the coupled network and define the error of the reduced model as

$$\mathscr{E}(\omega; P) := z^h(\omega) - z^{POD}(\omega; P), \tag{1.45}$$

where $z^h(\omega) := (e^h(\omega), j_V^h(\omega), j_L^h(\omega), y^h(\omega))^\top$ is the solution of Problem 1.3.1 at the parameter ω with discretized semiconductor variables $y^h := (\psi^h, n^h, p^h, g_\psi^h, J_n^h, J_p^h)^\top$. $z^{POD}(\omega; P)$ denotes the solution of the coupled system in Problem 1.4.1 with reduced semiconductors, where the reduced model is created based on simulations at the reference parameters $P \subset \mathscr{P}$. The error is considered in the space X with norm

$$\|z\|_X := \left\| \left(\|e\|_2, \|j_V\|_2, \|j_L\|_2, \right. \right.$$

$$\|\psi\|_{L^2([0,T], L^2(\Omega))}, \|n\|_{L^2([0,T], L^2(\Omega))}, \|p\|_{L^2([0,T], L^2(\Omega))},$$

$$\|g_\psi\|_{L^2([0,T], H_{0,N}(\mathrm{div}, \Omega))},$$

$$\left. \left. \|J_n\|_{L^2([0,T], H_{0,N}(\mathrm{div}, \Omega))}, \|J_p\|_{L^2([0,T], H_{0,N}(\mathrm{div}, \Omega))} \right) \right\|.$$

Obvious extensions apply when there is more than one semiconductor present.

Furthermore we define the residual $\mathscr{R}$ by evaluation of the unreduced model (1.31)–(1.35) at the solution of the reduced model $z^{POD}(\omega; P)$, i.e.

$$\mathscr{R}(z^{POD}(\omega; P)) := \begin{pmatrix} 0 \\ -M_L \dot{n}^{POD}(t) \\ M_L \dot{p}^{POD}(t) \\ 0 \\ 0 \\ 0 \end{pmatrix} + A_{FEM} \begin{pmatrix} \psi^{POD}(t) \\ n^{POD}(t) \\ p^{POD}(t) \\ g_\psi^{POD}(t) \\ J_n^{POD}(t) \\ J_p^{POD}(t) \end{pmatrix}$$

$$+ \mathscr{F}(n^{POD}, p^{POD}, g_\psi^{POD}) - b(A_S^T e^{POD}(t)). \tag{1.46}$$

Note that the residual of Eqs. (1.31)–(1.34) vanishes.

We note that the same definitions are used in [22] for linear descriptor systems. In [22] an error estimate is obtained by deriving a linear ODE for the error and exploiting explicit solution formulas. Here we have a nonlinear DAE and at the present state we are not able to provide an upper bound for the error $\|\mathscr{E}(\omega; P)\|_X$ which would yield a rigorous sampling method using for example the Greedy algorithm of [33].

We propose to consider the residual as an estimate for the error. The evaluation of the residual is cheap since it only requires the solution of the reduced system and its evaluation in the unreduced DAE. It is therefore possible to evaluate the residual at a large set of test parameters $P_{test} \subset \mathscr{P}$. Similar to the Greedy algorithm of [33], we add to the set of reference parameters the parameter where the residual becomes maximal.

The magnitude of the components in error and residual may be large and a proper scaling should be applied. For the error we consider the component-wise relative error, i.e.

$$\frac{\|\psi^h(\omega) - \psi^{POD}(\omega; P)\|_{L^2([0,T],L^2(\Omega))}}{\|\psi^h(\omega)\|_{L^2([0,T],L^2(\Omega))}}, \quad \frac{\|n^h(\omega) - n^{POD}(\omega; P)\|_{L^2([0,T],L^2(\Omega))}}{\|n^h(\omega)\|_{L^2([0,T],L^2(\Omega))}}, \quad \ldots,$$

and the residual is scaled by a block-diagonal matrix containing the weights

$$D(\omega)\mathscr{R}(z^{POD}(\omega; P)) = \begin{pmatrix} d_\psi(\omega)I & & & & & \\ & d_n(\omega)I & & & & \\ & & d_p(\omega)I & & & \\ & & & d_{g_\psi}(\omega)I & & \\ & & & & d_{J_n}(\omega)I & \\ & & & & & d_{J_p}(\omega)I \end{pmatrix}$$
$$\times \mathscr{R}(z^{POD}(\omega; P)).$$

The weights $d_{(\cdot)}(\omega) > 0$ may be parameter-dependent. These weights are chosen in a way that the norm of the residual and the relative error are component-wise equal at the reference frequencies ω_k where we know $z^h(\omega_k)$ from simulation of the unreduced model, i.e.

$$d_\psi(\omega_k) := \frac{\|\psi^h(\omega_k) - \psi^{POD}(\omega_k; P)\|_{L^2([0,T],L^2(\Omega))}}{\|\psi^h(\omega_k)\|_{L^2([0,T],L^2(\Omega))} \cdot \|\mathscr{R}_1(z^{POD}(\omega_k; P))\|_{L^2([0,T],L^2(\Omega))}}, \quad (1.47)$$

and similarly for the other components. If $\|\mathscr{R}_1(z^{POD}(\omega_k; P))\|_{L^2([0,T],L^2(\Omega))} = 0$ holds we chose $d_\psi(\omega_k) := 1$.

In one dimensional parameter sampling with $\mathscr{P} := [\underline{p}, \overline{p}]$, we approximate $d_{(\cdot)}(\omega)$ by piecewise linear interpolation of the weights $d_{(\cdot)}(\omega_1), \ldots, d_{(\cdot)}(\omega_k)$. Extrapolation is done by nearest-neighbor interpolation to ensure the positivity of the weights.

Algorithm 1.1 Sampling

---

1. Select $\omega_1 \in \mathscr{P}$, $P_{test} \subset \mathscr{P}$, $tol > 0$, and set $k := 1$, $P_1 := \{\omega_1\}$.
2. Simulate the unreduced model at ω_1 and calculate the reduced model with POD basis functions U_1.
3. Calculate weight functions $d_{(\cdot)}(\omega) > 0$ according to (1.47) for all $\omega_k \in P_k$.
4. Calculate the scaled residual $\|D(\omega)\mathscr{R}(z^{POD}(\omega, P_k))\|$ for all $\omega \in P_{test}$.
5. Check termination conditions, e.g.

   - $\max_{\omega \in P_{test}} \|D(\omega)\mathscr{R}(z^{POD}(\omega, P_k))\| < tol$,
   - no progress in weighted residual.

6. Calculate $\omega_{k+1} := \arg\max_{\omega \in P_{test}} \|D(\omega)\mathscr{R}(z^{POD}(\omega, P_k))\|$.
7. Simulate the unreduced model at ω_{k+1} and create a new reduced model with POD basis U_{k+1} using also the already available information at $\omega_1, \ldots, \omega_k$.
8. Set $P_{k+1} := P_k \cup \{\omega_{k+1}\}$, $k := k + 1$ and goto 3.

We summarize our ideas in the sampling Algorithm 1.1. The step 7 in this algorithm can be executed in different ways. If offline time and offline memory requirements are not critical one may combine snapshots from all simulations of the full model and redo the model order reduction on the large snapshot ensemble. Otherwise we can create a new reduced model at reference frequency ω_{k+1} with POD-basis $\bar{U}$ and then perform an additional POD step on $(U_k, \bar{U})$.

1.5.1 Numerical Investigation for Residual Based Sampling

We now apply Algorithm 1.1 to provide a reduced-order model of the basic circuit and we choose the frequency of the input voltage v_s as model parameter. As parameter space we chose the interval $\mathscr{P} := [10^8, 10^{12}]$ Hz. We start the investigation with a reduced model which is created from the simulation of the full model at the reference frequency $\omega_1 := 10^{10}$ Hz. The number of POD basis functions s is chosen such that the lack of information $\Delta(s)$ is approximately 10^{-7}. The relative error and the weighted residual are plotted in Fig. 1.12 (left). We observe that the weighted residual is a rough estimate for the relative approximation error. Using Algorithm 1.1 the next additional reference frequency is $\omega_2 := 10^8$ Hz since it maximizes the weighted residual. The second reduced model is constructed on the same lack of information $\Delta := 10^{-7}$. Here we note that in practical applications, the error is not known over the whole parameter space.

The next two iterations of the sampling algorithm are also depicted in Fig. 1.12. Based on the residual in step 2, one selects $\omega_3 := 1.0608 \cdot 10^9$ Hz as the next reference frequency. Since no further progress of the weighted residual is achieved in step 3, the algorithm terminates. The maximal errors and residuals are given in Table 1.3.

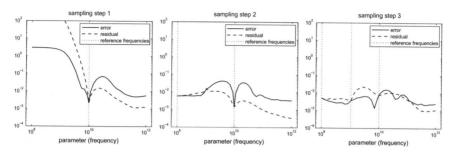

Fig. 1.12 *Left*: Relative reduction error (*solid line*) and weighted residual (*dashed line*) plotted over the frequency parameter space. The reduced model is created based on simulations at the reference frequency $\omega_1 := 10^{10}$ Hz, which is marked by *vertical dotted line*. *Middle*: Relative reduction error (*solid line*) and weighted residual (*dashed line*) plotted over the frequency parameter space. The reduced model is created based on simulations at the reference frequencies $\omega_1 := 10^{10}$ Hz and $\omega_2 := 10^8$ Hz. The reference frequencies are marked by *vertical dotted lines*. *Right*: Relative reduction error (*solid line*) and weighted residual (*dashed line*) plotted over the frequency parameter space. The reduced model is created based on simulations at the reference frequency $\omega_1 := 10^{10}$ Hz, $\omega_2 := 10^8$ Hz, and $\omega_3 := 1.0608 \cdot 10^9$ Hz. The reference frequencies are marked by *vertical dotted lines*

Table 1.3 Progress of refinement method

Step k	Reference parameters P_k	Max. scaled residual (at frequency)	Max. relative error (at frequency)
1	$\{1.0000 \cdot 10^{10}\}$	$9.9864 \cdot 10^2$ $(1.0000 \cdot 10^8)$	$3.2189 \cdot 10^0$ $(1.0000 \cdot 10^8)$
2	$\{1.0000 \cdot 10^8,$ $1.0000 \cdot 10^{10}\}$	$1.5982 \cdot 10^{-2}$ $(1.0608 \cdot 10^9)$	$4.3567 \cdot 10^{-2}$ $(3.4551 \cdot 10^9)$
3	$\{1.0000 \cdot 10^8,$ $1.0608 \cdot 10^9,$ $1.0000 \cdot 10^{10}\}$	$2.2829 \cdot 10^{-2}$ $(2.7283 \cdot 10^9)$	$1.6225 \cdot 10^{-2}$ $(1.8047 \cdot 10^{10})$

1.6 PABTEC Combined with POD MOR

In the current section, we combine the PABTEC approach of Chap. 2 and simulation based POD model order reduction techniques to determine reduced-order models for coupled circuit-device systems. While the PABTEC method preserves the passivity and reciprocity in the reduced linear circuit model, the POD approach delivers high-fidelity reduced-order models for the semiconductor devices. Details of the approach are given in [27].

Now we return to the network equations (1.31)–(1.35). The coupling relation (1.34) can shortly be written as $j_S(t) = \vartheta(x_S(t))$, where $x_S(t) = \left[\psi^T(t), n^T(t), p^T(t), g_\psi^T(t), J_n^T(t), J_p^T(t) \right]^T$ is the state vector of the semidiscretized drift-diffusion equations (1.35). Determining the state $x_S(t)$ from Eq. (1.35) for a given voltage $A_S^T e(t)$, say $x_S(t) = \chi(A_S^T e(t))$, and substituting it into (1.34), we obtain the relationship

$$j_S(t) = g(A_S^T e(t)), \tag{1.48}$$

where $g(A_S^T e(t)) := \vartheta(\chi(A_S^T e(t)))$ describes the voltage-current relation for the semidiscretized semiconductors. This relation can be considered as an input-to-output map, where the input is the voltage vector $A_S^T e(t)$ at the contacts of the semiconductors and the output is the approximate semiconductor current $j_S(t)$.

Electrical networks usually contains very large linear subnetworks modeling interconnects. In POD MOR we need to simulate the coupled DAE system (1.31)–(1.35) in order to determine the snapshots. To reduce the simulation time, we can first to separate the linear subsystem and approximate it by a reduced-order linear model of lower dimension using the PABTEC algorithm [38, 51], see also Chap. 2 in this book. The decoupled device equations are then reduced using the POD method presented in Sect. 1.4. Combining these reduced-order linear and nonlinear models, we obtain a nonlinear reduced-order model that approximates the coupled system (1.31)–(1.35).

1.6.1   Decoupling

For the extraction of a linear subcircuit, we use a decoupling procedure from [47] that consists in the replacement of the nonlinear inductors and nonlinear capacitors by controlled current sources and controlled voltage sources, respectively. The nonlinear resistors and semiconductor devices are replaced by an equivalent circuit consisting of two serial linear resistors and one controlled current source connected parallel to one of the resistors. Such replacements introduce additional nodes and state variables, but neither additional loops consisting of capacitors and voltage sources (CV-loops) nor cutsets consisting of inductors and current sources (LI-cutsets) occur in the decoupled linear subcircuit meaning that its index coincides with the index of the original circuit, see [13] for the index analysis of the circuit equations. An advantage of the suggested replacement strategy is demonstrated in the following example.

Example 1.6.1 Consider a circuit with a semiconductor diode as in Fig. 1.13. We suggest to replace the diode by an equivalent circuit shown in Fig. 1.14. If we would replace the diode by a current source, then a decoupled linear circuit would have I-cutset and, hence, lack well-posedness. Moreover, if we would replace the diode by a voltage source, then the resulting linear circuit would have CV-loop, i.e., it would be of index two, although the original circuit is of index one. Note that model reduction of index two problems is more involved than of index one problems [50].

For simplicity, we assume that the circuit does not contain nonlinear devices other than semiconductors. Then after the replacements described above, the extracted

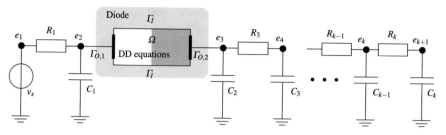

Fig. 1.13 RC chain with a diode

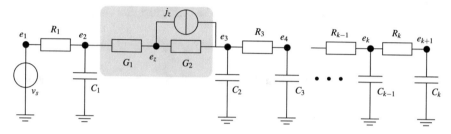

Fig. 1.14 Decoupled linear RC chain with a circuit replacing the diode

linear subcircuit can be modeled by the linear DAE system in the MNA form

$$E\dot{x}(t) = Ax(t) + Bu_l(t),\tag{1.49a}$$

$$y_l(t) = B^T x(t),\tag{1.49b}$$

with $x(t) = \left[\, e^T(t)\ e_z^T(t) \,\middle|\, j_L^T(t) \,\middle|\, j_V^T(t)\, \right]$, $u_l^T(t) = \left[\, i_s^T(t)\ j_z^T(t) \,\middle|\, v_s^T(t)\, \right]$ and

$$E = \begin{bmatrix} A_{C,l}CA_{C,l}^T & 0 & 0 \\ 0 & L & 0 \\ 0 & 0 & 0 \end{bmatrix}, \quad A = \begin{bmatrix} -A_{R,l}G_l A_{R,l}^T & -A_{L,l} & -A_{V,l} \\ A_{L,l}^T & 0 & 0 \\ A_{V,l}^T & 0 & 0 \end{bmatrix},$$

$$B = \begin{bmatrix} -A_{I,l} & 0 \\ 0 & 0 \\ 0 & -I \end{bmatrix},\tag{1.49c}$$

where the incidence and element matrices are given by

$$A_{C,l} = \begin{bmatrix} A_C \\ 0 \end{bmatrix}, \quad A_{L,l} = \begin{bmatrix} A_L \\ 0 \end{bmatrix}, \quad A_{V,l} = \begin{bmatrix} A_V \\ 0 \end{bmatrix}, \quad A_{I,l} = \begin{bmatrix} A_I & A_S^2 \\ 0 & I \end{bmatrix},\tag{1.49d}$$

$$A_{R,l} = \begin{bmatrix} A_R & A_S^1 & A_S^2 \\ 0 & -I & I \end{bmatrix}, \quad G_l = \begin{bmatrix} G & 0 & 0 \\ 0 & G_1 & 0 \\ 0 & 0 & G_2 \end{bmatrix}.\tag{1.49e}$$

Here, C, L and G are the capacitance, inductance and conductance matrices, A_S^1 and A_S^2 have entries in $\{0, 1\}$ and $\{-1, 0\}$, respectively, and satisfy $A_S^1 + A_S^2 = A_S$. Moreover, $e_z(t)$ is the potential of the introduced nodes, and the new input variable $j_z(t)$ is given by

$$j_z(t) = (G_1 + G_2)G_1^{-1}g(A_S^T e(t)) - G_2 A_S^T e(t), \tag{1.50}$$

where the matrices G_1 and G_2 are diagonal with conductances of the introduced linear resistors in the replacement circuits on the diagonal. One can show that the linear system (1.49) together with the decoupled nonlinear equations (1.35), (1.48) is state equivalent to the coupled system (1.31)–(1.35) together with the equation

$$e_z(t) = (G_1 + G_2)^{-1}\left(G_1(A_{\mathcal{R}}^1)^T e(t) - G_2(A_{\mathcal{R}}^2)^T e(t) - j_z(t)\right) \tag{1.51}$$

in the sense that these both systems have the same state vectors up to a permutation, see [47] for detail.

1.6.2 Model Reduction Approach

Applying the PABTEC method to the linear DAE system (1.49), we obtain a reduced-order model

$$\hat{E}\frac{d}{dt}\hat{x}(t) = \hat{A}\hat{x}(t) + \begin{bmatrix} \hat{B}_1 & \hat{B}_2 & \hat{B}_3 \end{bmatrix}\begin{bmatrix} i_s(t) \\ j_z(t) \\ v_s(t) \end{bmatrix}, \qquad \begin{bmatrix} \hat{y}_{l,1}(t) \\ \hat{y}_{l,2}(t) \\ \hat{y}_{l,3}(t) \end{bmatrix} = \begin{bmatrix} \hat{C}_1 \\ \hat{C}_2 \\ \hat{C}_3 \end{bmatrix}\hat{x}(t), \tag{1.52}$$

where $\hat{y}_{l,j} = \hat{C}_j\hat{x}(t)$, $j = 1, 2, 3$, approximate the corresponding components of the output y_l in (1.49b). Combining this reduced model with the semidiscretized drift-diffusion equations (1.35) via (1.48), we can determine the approximate snapshots which can then be used to compute the POD-reduced model as in (1.42). The coupling relation (1.41) can then be approximated by

$$\hat{j}_S(t) = C_1 U_{J_n}\gamma_{J_n}(t) + C_2 U_{J_p}\gamma_{J_p}(t) + C_3 U_{g_\psi}\dot{\gamma}_{g_\psi}(t). \tag{1.53}$$

As for the original system (1.34) and (1.35), we denote the relation between $A_S^T e(t)$ and $\hat{j}_S(t)$ by

$$\hat{j}_S(t) = \hat{g}(A_S^T e(t)). \tag{1.54}$$

Using (1.50) and (1.51), we have $-(A_S^2)^T e(t) - e_z(t) = -A_S^T e(t) + G_1 g(A_S^T e(t))$. Then it follows from $-(A_S^2)^T e(t) - e_z(t) \approx \hat{C}_2\hat{x}(t)$ that the semiconductor voltage vector $u_S(t) = A_S^T e(t)$ can be approximated by $\hat{u}_S(t)$ satisfying $-G_1 \hat{C}_2 \hat{x}(t) - G_1 \hat{u}_S(t) + \hat{g}(\hat{u}_S(t)) = 0$. Thus, combining the reduced linear system (1.52) with the reduced semiconductor model (1.42), we obtain a reduced-order coupled DAE system

$$\hat{E}\frac{d}{dt}\hat{x}(t) - (\hat{A} + \hat{B}_2(G_1 + G_2)\hat{C}_2)\hat{x}(t) - \hat{B}_2 G_1 \hat{u}_S(t) - \hat{B}_1 i_s(t) - \hat{B}_3 v_s(t) = 0,$$

$$(1.55)$$

$$-G_1 \hat{C}_2 \hat{x}(t) - G_1 \hat{u}_S(t) + \hat{g}(\hat{u}_S(t)) = 0,$$

$$(1.56)$$

$$\hat{j}_S(t) - C_1 U_{J_n} \gamma_{J_n}(t) - C_2 U_{J_p} \gamma_{J_p}(t) - C_3 U_{g_\psi} \dot{\gamma}_{g_\psi}(t) = 0,$$

$$(1.57)$$

$$\begin{pmatrix} 0 \\ -\dot{\gamma}_n(t) \\ \dot{\gamma}_p(t) \\ 0 \\ 0 \\ 0 \end{pmatrix} + A_{POD} \begin{pmatrix} \gamma_\psi(t) \\ \gamma_n(t) \\ \gamma_p(t) \\ \gamma_{g_\psi}(t) \\ \gamma_{J_n}(t) \\ \gamma_{J_p}(t) \end{pmatrix} + U^\top \mathscr{F}(n^{POD}, p^{POD}, g_\psi^{POD}) - U^\top b(\hat{u}_S(t)) = 0.$$

$$(1.58)$$

Note that model reduction of the linear subsystem and the semiconductor model can be executed independently.

1.6.3 Numerical Experiments

In this section, we present some results of numerical experiments to demonstrate the applicability of the presented model reduction approaches for coupled circuit-device systems.

For model reduction of linear circuit equations, we use the MATLAB Toolbox PABTEC, see Chap. 2. The POD method is implemented in C++ based on the FEM library deal.II [5] for discretizing the drift-diffusion equations. The obtained large and sparse nonlinear DAE system (1.31)–(1.35) as well as the small and dense reduced-order model (1.55)–(1.58) are integrated using the DASPK software package [9] based on a BDF method, where the nonlinear equations are solved using Newton's method. Furthermore, the direct sparse solver SuperLU [12] is employed for solving linear systems.

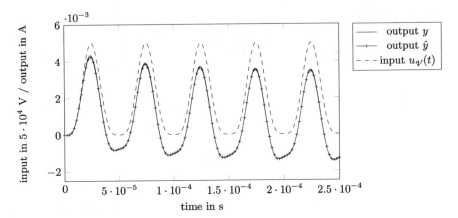

Fig. 1.15 Input voltage and output currents for the basic diode with the voltage-current relation (1.59)

Consider again an RC circuit with one diode as shown in Fig. 1.13. The input is given by

$$v_s(t) = u_{\mathcal{V}}(t) = 10\sin(2\pi f_0 t)^4$$

with the frequency $f_0 = 10^4$ Hz, see Fig. 1.15. The output of the system is $y(t) = -j_V(t)$. We simulate the models over the fixed time horizon $[0, \frac{2.5}{f_0}]$. The linear resistors have the same resistance $R = 2$ kΩ and the linear capacitors have the same capacitance $C = 0.02$ μF.

First, we describe the diode by the voltage-current relation

$$g(u_S) = 10^{-14}\left(\exp(40u_S) - 1\right), \tag{1.59}$$

and apply only the PABTEC method to the decoupled linear system (1.49) that models the linear circuit given in Fig. 1.14. System (1.49) with $n_l = 1503$ variables was approximated by a reduced model (1.52) of dimension 24. The outputs y and $\hat{y}$ of the original nonlinear system (1.31)–(1.33), (1.48), (1.59) and the reduced-order nonlinear model (1.55), (1.56) with $\hat{g}$ replaced by g are plotted in Fig. 1.15. Simulation time and the absolute and relative L_2-norm errors in the output are presented in Table 1.4. One can see that the simulation time is reduced by a factor of 10, while the relative error is below 2%.

As the next step, we introduce the drift-diffusion model (1.17)–(1.22) for the diode. The parameters of the diode are summarized in Table 1.5. Note that we do not expect to obtain the same output y as in the previous experiment. To achieve this, one would need to perform a parameter identification for the drift-diffusion model which is not done in this paper. In Table 1.6, we collect the numerical results for different model reduction strategies. The outputs of the systems with the reduced network and/or POD-reduced diode are compared to the full semidiscretized model (1.31)–

Table 1.4 Simulation time and approximation errors for the nonlinear RC circuit with the basic diode described by the voltage-current relation (1.59)

System	Dimension	Simulation time (s)	Absolute error $\|y - \hat{y}\|_{L_2}$	Relative error $\|y - \hat{y}\|_{L_2}/\|y\|_{L_2}$
Unreduced	1503	0.584		
Reduced	24	0.054	$5.441 \cdot 10^{-7}$	$1.760 \cdot 10^{-2}$

Table 1.5 Diode model parameters

Parameter	Value
ε	$1.03545 \cdot 10^{-12}$ F/cm
U_T	0.0259 V
n_0	$1.4 \cdot 10^{10}$ 1/cm^3
μ_n	1350 cm^2/(V s)
τ_n	$330 \cdot 10^{-9}$ s
μ_p	480 cm^2/(V s)
τ_p	$33 \cdot 10^{-9}$ s
Ω	$[0, l_1] \times [0, l_2] \times [0, l_3]$
l_1 (length)	10^{-4} cm
l_2 (width)	10^{-5} cm
l_3 (depth)	10^{-5} cm
$N(\xi), \, \xi_1 < l_1/2$	$-9.94 \cdot 10^{15}$ 1/cm^3
$N(\xi), \, \xi_1 \geq l_1/2$	$4.06 \cdot 10^{18}$ 1/cm^3
FEM-mesh	500 elements, refined at $\xi_1 = l_1/2$

Table 1.6 Statistics for model reduction of the coupled circuit-device system

Network (MNA equations)	Diode (DD equations)	Dim.	Simul. time (s)	Jacobian evaluations	Absolute error $\|y - \hat{y}\|_{L_2}$	Relative error $\|y - \hat{y}\|_{L_2}/\|y\|_{L_2}$
Unreduced	Unreduced	7510	23.37	20		
Reduced	Unreduced	6031	16.90	17	$2.165 \cdot 10^{-8}$	$7.335 \cdot 10^{-4}$
Unreduced	Reduced	1609	1.51	16	$2.952 \cdot 10^{-6}$	$1.000 \cdot 10^{-1}$
Reduced	Reduced	130	1.19	11	$2.954 \cdot 10^{-6}$	$1.000 \cdot 10^{-1}$

(1.35) with 7510 variables. First, we reduce the extracted linear network and do not modify the diode. This reduces the number of variables by about 20%, and the simulation time is reduced by 27%. It should also be noted that the reduced network is not only smaller but it is also easier to integrate for the DAE solver. An indicator for the computational complexity is the number of Jacobian evaluations or, equivalently, the number of LU decompositions required during integration.

Finally, we create a POD-reduced model (1.42) for the diode. The number of columns s_* of the projection matrices U_* is determined from the condition $\Delta_* \leq tol_{POD}$ with Δ_* defined in (1.36) and a tolerance $tol_{POD} = 10^{-6}$ for each component. We also apply the DEIM method for the reduction of nonlinearity evaluations in the drift-diffusion model. The resulting reduced-order model (1.42) for the diode is

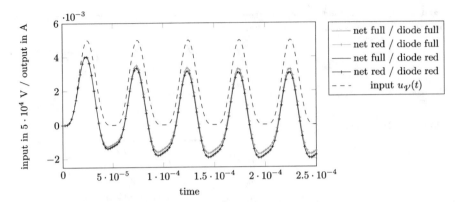

Fig. 1.16 Input voltage and output currents for the four model reduction setups

a dense DAE of dimension 105 while the original model (1.35) has dimension 6006, for the diode only. Coupling it with the unreduced and reduced linear networks, we obtain the results in Table 1.6 (last two rows). The simulation results for the different model reduction setups are also illustrated in Fig. 1.16.

The presented numerical results demonstrate that the recoupling of the respective reduced-order models delivers an overall reduced-order model for the circuit-device system which allows significantly faster simulations (speedup-factor is about 20) while keeping the relative errors below 10%.

Finally, we note that the model reduction concept developed in this section is not restricted to the reduction of electrical networks containing semiconductor devices. It can also be extended to the reduction of networks modeling e.g. nonlinear multibody systems containing many simple mass-spring-damper components and only a few high-fidelity components described by PDE systems.

Acknowledgements The work reported in this paper was supported by the German Federal Ministry of Education and Research (BMBF), grant no. 03HIPAE5. Responsibility for the contents of this publication rests with the authors.

References

1. Anile, A.M, Mascali, G., Romano, V.: Mathematical Problems in Semiconductor Physics. Lecture Notes in Mathematics. Springer, Berlin (2003). Lectures given at the C. I. M. E. Summer School, Cetraro, Italy, 15–22 July 1998
2. Anile, A., Nikiforakis, N., Romano, V., Russo, G.: Discretization of semiconductor device problems. II. In: Schilders, W.H.A., et al. (eds.) Special Volume: Numerical Methods in Electromagnetics. Handbook of Numerical Analysis, vol. 13. Elsevier/North Holland, Amsterdam; Handb. Numer. Anal. **13**, 443–522 (2005)
3. Antoulas, A., Beattie, C., Gugercin, S.: Interpolatory model reduction of large-scale dynamical systems. In: Mohammadpour, J., Grigoriadis, K. (eds.) Efficient Modeling and Control of Large-Scale Systems, pp. 3–58. Springer, New York (2010)

4. Astrid, P., Weiland, S., Willcox, K., Backx, T.: Missing point estimation in models described by proper orthogonal decomposition. IEEE Trans. Autom. Control **53**(10), 2237–2251 (2008)
5. Bangerth, W., Hartmann, R., Kanschat, G.: deal.II – a general-purpose object-oriented finite element library. ACM Trans. Math. Softw. **33**(4), Article No. 24 (2007)
6. Benner, P., Hinze, M., ter Maten, E. (eds.): Model Reduction for Circuit Simulation. Lecture Notes in Electrical Engineering, vol. 74. Springer, Berlin/Heidelberg (2011)
7. Bodestedt, M., Tischendorf, C.: PDAE models of integrated circuits and index analysis. Math. Comput. Model. Dyn. Syst. **13**(1), 1–17 (2007)
8. Brezzi, F., Marini, L., Micheletti, S., Pietra, P., Sacco, R., Wang, S.: Discretization of semiconductor device problems. I. In: Schilders, W., ter Maten, E. (eds.) Special Volume: Numerical Methods in Electromagnetics. Handbook of Numerical Analysis, vol. 13, pp. 317–441. Elsevier, Amsterdam (2005)
9. Brown, P., Hindmarsh, A., Petzold, A.: A description of DASPK: a solver for large-scale differential-algebraic systems. Lawrence Livermore National Report, UCRL (1992)
10. Chaturantabut, S., Sorensen, D.C.: Discrete empirical interpolation for nonlinear model reduction. Tech. Rep. 09-05, Department of Computational and Applied Mathematics, Rice University (2009)
11. Chaturantabut, S., Sorensen, D.: Nonlinear model reduction via discrete empirical interpolation. SIAM J. Sci. Comput. **32**(5), 2737–2764 (2010)
12. Demmel, J., Eisenstat, S., Gilbert, J., Li, X., Liu, J.: A supernodal approach to sparse partial pivoting. SIAM J. Matrix Anal. Appl. **20**(3), 720–755 (1999)
13. Estévez, S.D., Tischendorf, C.: Structural analysis for electric circuits and consequences for MNA. Int. J. Circuit Theory Appl. **28**, 131–162 (2000)
14. Estévez, S.D., Feldmann, U., März, R., Sturtzel, S., Tischendorf, C.: Finding beneficial DAE structures in circuit simulation. In: Jäger, W., et al. (eds.) Mathematics - Key Technology for the Future. Joint Projects Between Universities and Industry, pp. 413–428. Springer, Berlin (2003)
15. Freund, R.: SPRIM: structure-preserving reduced-order interconnect macromodeling. In: Technical Digest of the 2004 IEEE/ACM International Conference on Computer-Aided Design, pp. 80–87. IEEE Computer Society, Los Alamos, CA (2004)
16. Freund, R.: Structure-preserving model order reduction of RCL circuit equations. In: Schilders, W., van der Vorst, H.A., Rommes, J. (eds.) Model Order Reduction: Theory, Research Aspects and Applications. Mathematics in Industry, vol. 13, pp. 49–73. Springer, Berlin/Heidelberg (2008)
17. Günther, M.: Partielle differential-algebraische Systeme in der numerischen Zeitbereichs-analyse elektrischer Schaltungen. VDI Fortschritts-Berichte, Reihe 20, Rechnerunterstützte Verfahren, Nr. 343 (2001)
18. Günther, M., Feldmann, U., ter Maten, J.: Modelling and discretization of circuit problems. In: Schilders, W.H.A., et al. (eds.) Handbook of Numerical Analysis, vol. 13. Elsevier/North Holland, Amsterdam; Handb. Numer. Anal. **13**, 523–629 (2005)
19. Gugercin, S., Antoulas, A.: A survey of model reduction by balanced truncation and some new results. Int. J. Control **77**(8), 748–766 (2004)
20. Gugercin, S., Antoulas, A., Beattie, C.: $\mathscr{H}_2$ model reduction for large-scale linear dynamical systems. SIAM J. Matrix Anal. Appl. **30**(2), 609–638 (2008)
21. Gummel, H.: A self-consistent iterative scheme for one-dimensional steady state transistor calcula- tions. IEEE Trans. Electron Devices **11**(10), 455–465 (1964)
22. Haasdonk, B., Ohlberger, M.: Efficient reduced models and a-posteriori error estimation for parametrized dynamical systems by offline/online decomposition. SimTech Preprint 2009-23 (2009)
23. Hinze, M., Kunkel, M.: Residual based sampling in POD model order reduction of drift-diffusion equations in parametrized electrical networks. Z. Angew. Math. Mech. **92**, 91–104 (2012)
24. Hinze, M., Kunkel, M.: Discrete empirical interpolation in POD model order reduction of drift-diffusion equations in electrical networks. In: Scientific Computing in Electrical Engineering

SCEE 2010. Mathematics in Industry, vol. 16, Part 5, pp. 423–431. Springer, Berlin/Heidelberg (2012). doi:10.1007/978-3-642-22453-945

25. Hinze, M., Kunkel, M., Vierling, M.: POD model order reduction of drift-diffusion equations in electrical networks. In: Benner, P., Hinze, M., ter Maten, E. (eds.) Model Reduction for Circuit Simulation. Lecture Notes in Electrical Engineering, vol. 74, pp. 171–186. Springer, Berlin/Heidelberg (2011)

26. Hinze, M., Kunkel, M., Matthes, U.: POD model order reduction of electrical networks with semiconductors modeled by the transient drift–diffusion equations. In: Günther, M., Bartel, A., Brunk, M., Schöps, S., Striebel, M. (eds.) Progress in Industrial Mathematics at ECMI 2010. Mathematics in Industry, vol. 17, pp. 163–169. Springer, Berlin/Heidelberg (2012)

27. Hinze, M., Kunkel, M., Steinbrecher, A., Stykel, T.: Model order reduction of coupled circuit-device systems. Int. J. Numer. Model. **25**, 362–377 (2012)

28. Ho, C., Ruehli, A., Brennan, P.: The modified nodal approach to network analysis. IEEE Trans. Circuits Syst. **22**, 504–509 (1975)

29. Holmes, P., Lumley, J., Berkooz, G.: Turbulence, Coherent Structures, Dynamical Systems and Symmetry. Cambridge Monographs on Mechanics. Cambridge University Press, Cambridge (1996)

30. Markowich, P.: The Stationary Semiconductor Device Equations. Computational Microelectronics. Springer, Wien/New York (1986)

31. Moore, B.: Principal component analysis in linear systems: controllability, observability, and model reduction. IEEE Trans. Autom. Control **26**(1), 17–32 (1981)

32. Odabasioglu, A., Celik, M., Pileggi, L.: PRIMA: passive reduced-order interconnect macro-modeling algorithm. IEEE Trans. Comput. Aided Des. Integr. Circuits Syst. **17**(8), 645–654 (1998)

33. Patera, A., Rozza, G.: Reduced Basis Approximation and a Posteriori Error Estimation for Parametrized Partial Differential Equations. MIT Pappalardo Graduate Monographs in Mechanical Engineering. Massachusetts Institute of Technology, Cambridge (2007)

34. Petzold, L.R.: A description of DASSL: a differential/algebraic system solver. IMACS Trans. Sci. Comput. **1**, 65–68 (1993)

35. Phillips, J., Daniel, L., Silveira, L.: Guaranteed passive balancing transformations for model order reduction. IEEE Trans. Comput. Aided Des. Integr. Circuits Syst. **22**(8), 1027–1041 (2003)

36. Pinnau, R.: Model reduction via proper orthogonal decomposition. In: Schilders, W.H.A., et al. (eds.) Model Order Reduction: Theory, Research Aspects and Applications. Selected Papers Based on the Presentations at the Workshop 'Model Order Reduction, Coupled Problems and Optimization', Leiden, The Netherlands, 19–23 September 2005. Mathematics in Industry, vol. 13, pp. 95–109. Springer, Berlin (2008)

37. Reis, T., Stykel, T.: Positive real and bounded real balancing for model reduction of descriptor systems. Int. J. Control **83**(1), 74–88 (2010)

38. Reis, T., Stykel, T.: PABTEC: passivity-preserving balanced truncation for electrical circuits. IEEE Trans. Comput. Aided Des. Integr. Circuits Syst. **29**(9), 1354–1367 (2010)

39. Reis, T., Stykel, T.: Lyapunov balancing for passivity-preserving model reduction of RC circuits. SIAM J. Appl. Dyn. Syst. **10**(1), 1–34 (2011)

40. Rewieński, M.: A Trajectory Piecewise-Linear Approach to Model Order Reduction of Nonlinear Dynamical Systems. Ph.D. thesis, Massachusetts Institute of Technology (2003)

41. Rowley, C.: Model reduction for fluids, using balanced proper orthogonal decomposition. Int. J. Bifur. Chaos Appl. Sci. Eng. **15**(3), 997–1013 (2005)

42. Schilders, W., van der Vorst, H., Rommes, J. (eds.): Model Order Reduction: Theory, Research Aspects and Applications. Mathematics in Industry, vol. 13. Springer, Berlin/Heidelberg (2008)

43. Selberherr, S.: Analysis and Simulation of Semiconductor Devices. Springer, Wien/New York (1984)

44. Selva, S.M.: A coupled system for electrical circuits. Numerical simulations. PAMM **6**(1), 51–54 (2006)

45. Sirovich, L.: Turbulence and the dynamics of coherent structures. I: coherent structures. II: symmetries and transformations. III: dynamics and scaling. Q. Appl. Math. **45**, 561–590 (1987)
46. Soto, M.S., Tischendorf, C.: Numerical analysis of DAEs from coupled circuit and semiconductor simulation. Appl. Numer. Math. **53**(2–4), 471–88 (2005)
47. Steinbrecher, A., Stykel, T.: Model order reduction of nonlinear circuit equations. Int. J. Circuit Theory Appl. **41**, 1226–1247 (2013). doi:10.1002/cta.1821
48. Striebel, M., Rommes, J.: Model order reduction of nonlinear systems in circuit simulation: status and applications. In: Benner, P., Hinze, M., ter Maten, E. (eds.) Model Reduction for Circuit Simulation. Lecture Notes in Electrical Engineering, vol. 74, pp. 279–292. Springer, Berlin/Heidelberg (2011)
49. Stykel, T.: Gramian-based model reduction for descriptor systems. Math. Control Signals Syst. **16**, 297–319 (2004)
50. Stykel, T.: Balancing-related model reduction of circuit equations using topological structure. In: Benner, P., Hinze, M., ter Maten, E. (eds.) Model Reduction for Circuit Simulation. Lecture Notes in Electrical Engineering, vol. 74, pp. 53–80. Springer, Berlin/Heidelberg (2011)
51. Stykel, T., Reis, T.: The PABTEC algorithm for passivity-preserving model reduction of circuit equations. In: Proceedings of the 19th International Symposium on Mathematical Theory of Networks and Systems, MTNS 2010, Budapest, 5–9 July 2010, paper 363. ELTE, Budapest (2010)
52. Tischendorf, C.: Coupled Systems of Differential Algebraic and Partial Differential Equations in Circuit and Device Simulation. Habilitation thesis, Humboldt-Universität Berlin (2004)
53. Verhoeven, A., ter Maten, E., Striebel, M., Mattheij, R.: Model order reduction for nonlinear IC models. In: Korytowski, A., Malanowski, K., Mitkowski, W., Szymkat, M. (eds.) System Modeling and Optimization. IFIP Advances in Information and Communication Technology, vol. 312, pp. 476–491. Springer, Berlin/Heidelberg (2009)
54. Willcox, K., Peraire, J.: Balanced model reduction via the proper orthogonal decomposition. AIAA J. **40**(11), 2323–2330 (2002)

Chapter 2
Element-Based Model Reduction in Circuit Simulation

Andreas Steinbrecher and Tatjana Stykel

Abstract In this paper, we consider model reduction of linear and nonlinear differential-algebraic equations arising in circuit simulation. Circuit equations obtained using modified nodal or loop analysis have a special structure that can be exploited to construct efficient model reduction algorithms. For linear systems, we review passivity-preserving balanced truncation model reduction methods that are based on solving projected Lur'e or Lyapunov matrix equations. Furthermore, a topology-based index-preserving procedure for extracting large linear subnetworks from nonlinear circuits is given. Finally, we describe a new MATLAB Toolbox PABTEC for model reduction of circuit equations and present some results of numerical experiments.

2.1 Introduction

As integrated circuits get more complex and different physical effects have to be taken into account, the development of efficient modeling and simulation tools for very large networks is highly required. In this context, model order reduction is of crucial importance, especially if simulation of large-scale systems has to be done in a short time or it has to be repeated for different input signals. A general idea of model order reduction is to approximate a large-scale dynamical system by a reduced-order model that preserves essential properties like stability and passivity. It is also required that the approximation error is small.

Many different model reduction approaches have been developed in computational fluid dynamics, control design and electrical and mechanical engineering, see [3, 13, 61, 64] for books on this topic. One of the most used model reduction techniques in circuit simulation is *moment matching approximation* based on Krylov

A. Steinbrecher

Institut für Mathematik, Technische Universität Berlin, MA 4-5, Straße des 17. Juni 136, 10623 Berlin, Germany

e-mail: steinbrecher@math.tu-berlin.de

T. Stykel (✉)

Institut für Mathematik, Universität Augsburg, Universitätsstraße 14, 86159 Augsburg, Germany

e-mail: stykel@math.uni-augsburg.de

© Springer International Publishing AG 2017

P. Benner (ed.), *System Reduction for Nanoscale IC Design*,
Mathematics in Industry 20, DOI 10.1007/978-3-319-07236-4_2

subspace methods, e.g., [6, 25, 30]. Although these methods are efficient for very large sparse problems, the resulting reduced-order systems have only locally good approximation properties, and stability and passivity are not necessarily preserved. Furthermore, passivity-preserving model reduction methods based on Krylov subspaces have been developed for structured systems arising in circuit simulation [26, 27, 42, 48] and also for general systems [4, 38, 66]. However, none of these methods provides computable global error bounds. Another drawback of Krylov subspace methods is the *ad hoc* choice of interpolation points that strongly influence the approximation quality. An optimal point selection strategy based on tangential interpolation has been presented in [5, 32] that provides an optimal $\mathcal{H}_2$-approximation.

In this paper, we present a survey on passivity-preserving balanced truncation model reduction methods for linear circuit equations developed in [54, 56, 72]. They involve computing the spectral projectors onto the left and right deflating subspaces corresponding to the finite and infinite eigenvalues of an underlying pencil and solving projected matrix equations. An important property of these methods is the existence of computable error bounds that allow an adaptive choice of the order of the approximate model.

Furthermore, we consider model reduction of nonlinear circuits based on decoupling linear and nonlinear subcircuits followed by reduction of the linear part [68]. This model reduction approach can also be combined with the POD-based reduction technique for semiconductor devices, see Chap. 1, and further with hierarchical reduction methods studied in Chap. 4. The developed model reduction algorithms for circuit equations were implemented as MATLAB toolbox PABTEC and tested on practical problems.

Notation Throughout the paper, $\mathbb{R}^{n,m}$ and $\mathbb{C}^{n,m}$ denote the spaces of $n \times m$ real and complex matrices, respectively. The open left and right half-planes are denoted by $\mathbb{C}_-$ and $\mathbb{C}_+$, respectively, and $i = \sqrt{-1}$. The matrices A^T and A^* denote, respectively, the transpose and the conjugate transpose of $A \in \mathbb{C}^{n,m}$, and $A^{-T} = (A^{-1})^T$. An identity matrix of order n is denoted by I_n or simply by I. We use rank(A), im(A) and ker(A) for the rank, the range and the kernel of A, respectively. A matrix $A \in \mathbb{C}^{n,n}$ is *positive definite* (*semidefinite*), if $v^*Av > 0$ ($v^*Av \geq 0$) for all non-zero $v \in \mathbb{C}^n$. Note that positive (semi)definiteness of A does not require A to be Hermitian. For $A, B \in \mathbb{C}^{n,n}$, we write $A > B$ ($A \geq B$) if $A - B$ is positive definite (semidefinite). Furthermore, diag($A_1, \ldots, A_s$) denotes a block diagonal matrix with block entries $A_j, j = 1, \ldots, s$, on the diagonal.

2.2 Circuit Equations

In this section, we briefly describe the modeling of electrical circuits via differential-algebraic equations (DAEs) and discuss their properties. For more details on graph theory and network analysis, we refer to [1, 20, 22, 40, 75].

2.2.1 Graph-Theoretic Concepts

A general circuit can be modeled as a *directed graph* $\mathfrak{G}$ whose vertices (nodes) $\mathfrak{n}_k$ correspond to the nodes of the circuit and whose branches (edges) $\mathfrak{b}_{k_1,k_2} = \langle \mathfrak{n}_{k_1}, \mathfrak{n}_{k_2} \rangle$ correspond to the circuit elements like capacitors, inductors and resistors. For the ordered pair $\mathfrak{b}_{k_1,k_2} = \langle \mathfrak{n}_{k_1}, \mathfrak{n}_{k_2} \rangle$, we say that $\mathfrak{b}_{k_1,k_2}$ leaves $\mathfrak{n}_{k_1}$ and enters $\mathfrak{n}_{k_2}$. In this case, $\mathfrak{b}_{k_1,k_2}$ is called *incident* with $\mathfrak{n}_{k_1}$ and $\mathfrak{n}_{k_2}$. An alternating sequence $(\mathfrak{n}_{k_1}, \mathfrak{b}_{k_1,k_2}, \mathfrak{n}_{k_2}, \ldots, \mathfrak{n}_{k_{s-1}}, \mathfrak{b}_{k_{s-1},k_s}, \mathfrak{n}_{k_s})$ of vertices and branches in $\mathfrak{G}$ is called a *path* connecting $\mathfrak{n}_{k_1}$ and $\mathfrak{n}_{k_s}$ if the branches $\mathfrak{b}_{k_{j-1},k_j}$ are incident with the vertices $\mathfrak{n}_{k_{j-1}}$ and $\mathfrak{n}_{k_j}$ for $2 \leq j \leq s$. A path is *closed* if $\mathfrak{n}_{k_1} = \mathfrak{n}_{k_s}$. A closed path is called a *loop* if $\mathfrak{n}_{k_i} \neq \mathfrak{n}_{k_j}$ for $1 \leq i < j \leq s$ except for $\mathfrak{n}_{k_1}$ and $\mathfrak{n}_{k_s}$. A graph $\mathfrak{G}$ is called *connected* if for every two vertices there exists a path connecting them. A *cutset* is a set of branches of a connected graph whose removal disconnects the graph, and this set is minimal with this property. A subgraph of the graph $\mathfrak{G}$ is called a *tree* if it has all nodes of $\mathfrak{G}$, is connected and does not contain loops.

A directed graph $\mathfrak{G}$ with n_v vertices, n_b branches and n_l loops can be described by an *incidence matrix* $\mathbf{A}_0 = [a_{pq}] \in \mathbb{R}^{n_v, n_b}$ with

$$
a_{pq} = \begin{cases} 1 & \text{if branch } q \text{ leaves vertex } p, \\ -1 & \text{if branch } q \text{ enters vertex } p, \\ 0 & \text{otherwise,} \end{cases}
$$

or by a *loop matrix* $\mathbf{B}_0 = [b_{pq}] \in \mathbb{R}^{n_l, n_b}$ with

$$
b_{pq} = \begin{cases} 1 & \text{if branch } q \text{ belongs to loop } p \text{ and has the same orientation,} \\ -1 & \text{if branch } q \text{ belongs to loop } p \text{ and has the contrary orientation,} \\ 0 & \text{otherwise.} \end{cases}
$$

For a connected graph, the matrices $\mathbf{A}_0$ and $\mathbf{B}_0$ satisfy the following relations

$$
\ker(\mathbf{B}_0) = \operatorname{im}(\mathbf{A}_0^T), \qquad \operatorname{rank}(\mathbf{A}_0) = n_v - 1, \qquad \operatorname{rank}(\mathbf{B}_0) = n_b - n_v + 1,
$$

see [22, p. 213]. Removing linear dependent rows from $\mathbf{A}_0$ and $\mathbf{B}_0$, we obtain the full rank matrices $\mathbf{A} \in \mathbb{R}^{n_v-1, n_b}$ and $\mathbf{B} \in \mathbb{R}^{n_b-n_v+1, n_b}$ which are called the *reduced incidence matrix* and the *reduced loop matrix*, respectively.

2.2.2 Modified Nodal Analysis and Modified Loop Analysis

We now consider a general nonlinear RLC circuit that contains $n_{\mathcal{R}}$ resistors, n_L inductors, n_C capacitors, $n_{\mathcal{V}}$ independent voltage sources and $n_{\mathcal{J}}$ independent current sources. Such circuits are completely described by the graph-theoretic relations like Kirchhoff's current and voltage laws together with the branch constitutive relations that characterize the circuit elements. *Kirchhoff's current law* states that the sum of the currents along all branches leaving and entering a circuit node is zero. *Kirchhoff's voltage law* states that the sum of the voltages along the branches of any loop is zero. Let

$$j = [\, j_{\mathcal{R}}^T, \; j_C^T, \; j_L^T, \; j_{\mathcal{V}}^T, \; j_{\mathcal{J}}^T \,]^T \in \mathbb{R}^{n_b}, \qquad v = [\, v_{\mathcal{R}}^T, \; v_C^T, \; v_L^T, \; v_{\mathcal{V}}^T, \; v_{\mathcal{J}}^T \,]^T \in \mathbb{R}^{n_b}$$

denote the vectors of branch currents and branch voltages, respectively, and let the reduced incidence and loop matrices

$$\mathbf{A} = [\, A_{\mathcal{R}}, \; A_C, \; A_L, \; A_{\mathcal{V}}, \; A_{\mathcal{J}} \,], \qquad\qquad \mathbf{B} = [\, B_{\mathcal{R}}, \; B_C, \; B_L, \; B_{\mathcal{V}}, \; B_{\mathcal{J}} \,]$$

be partitioned accordingly, where the subscripts $\mathcal{R}$, C, L, $\mathcal{V}$ and $\mathcal{J}$ stand for resistors, capacitors, inductors, voltage sources and current sources, respectively. Then Kirchhoff's current and voltage laws can be expressed in the compact form as

$$\mathbf{A} j = 0, \qquad\qquad \mathbf{B} v = 0,$$

respectively, or, equivalently,

$$\mathbf{B}^T \iota = j, \qquad\qquad \mathbf{A}^T \eta = v,$$

where $\iota \in \mathbb{R}^{n_b - n_v + 1}$ and $\eta \in \mathbb{R}^{n_v - 1}$ denote the vectors of loop currents and node potentials.

The *branch constitutive relations* for nonlinear capacitors, inductors and resistors are given by

$$\tfrac{d}{dt}\phi(j_L) = v_L, \qquad j_C = \tfrac{d}{dt} q_C(v_C), \qquad j_{\mathcal{R}} = g(v_{\mathcal{R}}), \qquad\qquad (2.1)$$

where the functions $\phi : \mathbb{R}^{n_L} \to \mathbb{R}^{n_L}$, $q_C : \mathbb{R}^{n_C} \to \mathbb{R}^{n_C}$ and $g : \mathbb{R}^{n_{\mathcal{R}}} \to \mathbb{R}^{n_{\mathcal{R}}}$ describe electromagnetic fluxes in the inductors, capacitor charges and resistor voltage-current characteristics, respectively. For current-controlled resistors, we have also the relation $v_{\mathcal{R}} = \varrho(j_{\mathcal{R}})$, where $\varrho : \mathbb{R}^{n_{\mathcal{R}}} \to \mathbb{R}^{n_{\mathcal{R}}}$ is the resistor current-voltage characteristic function. We assume that

(A1) the functions ϕ, q_C and g are continuously differentiable and their Jacobians

$$\frac{\partial \phi(j_L)}{\partial j_L} = L(j_L), \qquad \frac{\partial q_C(v_C)}{\partial v_C} = C(v_C), \qquad \frac{\partial g(v_{\mathcal{R}})}{\partial v_{\mathcal{R}}} = G(v_{\mathcal{R}}),$$

are positive definite for all j_L, v_C and $v_{\mathcal{R}}$, respectively.

This assumption guarantees that inductors, capacitors and resistors are *locally passive*, see [19] for details.

Using Kirchhoff's laws and the branch constitutive relations, the dynamical behaviour of a nonlinear circuit can be described using *modified nodal analysis* (MNA) [37] by the following system of DAEs

$$\mathscr{E}(x)\tfrac{d}{dt}x = \mathscr{A}x + f(x) + \mathscr{B}u,$$
$$y = \mathscr{B}^T x, \tag{2.2}$$

where

$$\mathscr{E}(x) = \begin{bmatrix} A_C C(A_C^T \eta) A_C^T & 0 & 0 \\ 0 & L(j_L) & 0 \\ 0 & 0 & 0 \end{bmatrix}, \quad \mathscr{A} = \begin{bmatrix} 0 & -A_L & -A_V \\ A_L^T & 0 & 0 \\ A_V^T & 0 & 0 \end{bmatrix},$$

$$f(x) = \begin{bmatrix} -A_R g(A_R^T \eta) \\ 0 \\ 0 \end{bmatrix}, \quad \mathscr{B} = \begin{bmatrix} -A_J & 0 \\ 0 & 0 \\ 0 & -I \end{bmatrix}, \quad x = \begin{bmatrix} \eta \\ j_L \\ j_V \end{bmatrix},$$

$$\tag{2.3}$$

and the input u and the output y have the form

$$u = \begin{bmatrix} j_J \\ v_V \end{bmatrix}, \qquad y = \begin{bmatrix} -v_J \\ -j_V \end{bmatrix}, \tag{2.4}$$

respectively. Another approach for modeling electrical circuits is based on *modified loop analysis* (MLA), see [79]. In this case, the circuit equations take the form (2.2) with

$$\mathscr{E}(x) = \begin{bmatrix} B_L L(B_L^T \iota) B_L^T & 0 & 0 \\ 0 & C(v_C) & 0 \\ 0 & 0 & 0 \end{bmatrix}, \quad \mathscr{A} = \begin{bmatrix} 0 & -B_C & -B_J \\ B_C^T & 0 & 0 \\ B_J^T & 0 & 0 \end{bmatrix},$$

$$f(x) = \begin{bmatrix} -B_R \varrho(B_R^T \iota) \\ 0 \\ 0 \end{bmatrix}, \quad \mathscr{B} = \begin{bmatrix} 0 & -B_V \\ 0 & 0 \\ -I & 0 \end{bmatrix}, \quad x = \begin{bmatrix} \iota \\ v_C \\ v_J \end{bmatrix},$$

$$\tag{2.5}$$

and the input and the output are as in (2.4).

We assume that the circuit is well-posed in the sense that

(A2) the circuit does not contain cutsets consisting of current sources (I-cutsets),
(A3) the circuit does not contain loops consisting of voltage sources (V-loops).

These assumptions avoid open-circuit current sources and short-circuit voltage sources, respectively. Assumption (A2) is equivalent to

$$\text{rank}([A_C, A_L, A_R, A_V]) = n_v - 1,$$

which is, on the other hand, equivalent to $\text{rank}(B_{\mathcal{J}}) = n_{\mathcal{J}}$. In terms of rank conditions, (A3) means that $\text{rank}(A_{\mathcal{V}}) = n_{\mathcal{V}}$ or, equivalently, $\text{rank}([B_C, B_L, B_{\mathcal{R}}, B_{\mathcal{J}}]) = n_b - n_v + 1$.

The index concept plays an important role in the analysis of DAEs. To characterize different analytical and numerical properties of DAE systems, several index notations have been introduced in the literature, e.g., [17, 29, 33, 43]. For example, the *differentiation index* is roughly defined as the minimum of times that all or part of a DAE system must be differentiated with respect to t in order to determine the derivative of x as a continuous function of t and x. In the sequel, we will use the shorter term "index" instead of "differentiation index". The following proposition characterizes the index of the MNA system (2.2), (2.3).

Proposition 2.2.1 ([24, 68]) *Consider a circuit satisfying assumptions* (A1)–(A3).

1. *The index of the MNA system* (2.2), (2.3) *is at most two.*
2. *The index of the MNA system* (2.2), (2.3) *is equal to zero if and only if*

$$n_{\mathcal{V}} = 0 \quad and \quad \text{rank}(A_C) = n_v - 1. \tag{2.6}$$

3. *The index of the MNA system* (2.2), (2.3) *is equal to one if and only if*

$$\text{rank}([A_C, A_{\mathcal{V}}]) = \text{rank}(A_C) + n_{\mathcal{V}} \quad and \quad \text{rank}([A_C, A_{\mathcal{R}}, A_{\mathcal{V}}]) = n_v - 1. \tag{2.7}$$

Similar, rank conditions can also be formulated for the MLA system (2.2), (2.5). Considering the topological structure of the circuit, the conditions (2.6) imply that the circuit does not contain voltage sources and the circuit graph contains a capacitive tree, respectively. Furthermore, the first condition in (2.7) implies that the circuit does not contain loops consisting of capacitors and/or voltage sources (CV-loops) except for loops consisting of capacitors only (C-loops), whereas the second condition in (2.7) means that the circuit does not contain cutsets consisting of inductors and/or current sources (LI-cutsets).

In the following, we will distinguish between nonlinear circuits, which contain nonlinear elements, and linear circuits consisting exclusively of linear capacitors, inductors and resistors. A circuit element is called linear if the current-voltage relation for this element is linear. Otherwise, the circuit element is called nonlinear. Without loss of generality we may assume that the circuit elements are ordered such that the incidence matrices are partitioned as

$$A_C = \begin{bmatrix} A_{\tilde{C}}, & A_{\tilde{C}} \end{bmatrix}, \quad A_L = \begin{bmatrix} A_{\tilde{L}}, & A_{\tilde{L}} \end{bmatrix}, \quad A_{\mathcal{R}} = \begin{bmatrix} A_{\tilde{\mathcal{R}}}, & A_{\tilde{\mathcal{R}}} \end{bmatrix}, \tag{2.8}$$

where the incidence matrices $A_{\tilde{C}}$, $A_{\tilde{L}}$ and $A_{\tilde{\mathcal{R}}}$ correspond to the linear circuit components, and $A_{\tilde{C}}$, $A_{\tilde{L}}$ and $A_{\tilde{\mathcal{R}}}$ are the incidence matrices for the nonlinear devices. We also assume that the linear and nonlinear elements are not mutually connected, i.e.,

$$C(A_C^T\eta) = \begin{bmatrix} \bar{C} & 0 \\ 0 & \widetilde{C}(A_{\widetilde{C}}^T\eta) \end{bmatrix}, \qquad L(j_L) = \begin{bmatrix} \bar{L} & 0 \\ 0 & \widetilde{L}(j_{\widetilde{L}}) \end{bmatrix}, \qquad g(A_{\mathcal{R}}^T\eta) = \begin{bmatrix} \bar{G}A_{\bar{\mathcal{R}}}^T\eta \\ \widetilde{g}(A_{\widetilde{\mathcal{R}}}^T\eta) \end{bmatrix},$$

where $\bar{C} \in \mathbb{R}^{n_{\bar{C}},n_{\bar{C}}}$, $\bar{L} \in \mathbb{R}^{n_{\bar{L}},n_{\bar{L}}}$ and $\bar{G} \in \mathbb{R}^{n_{\bar{\mathcal{R}}},n_{\bar{\mathcal{R}}}}$ are the capacitance, inductance and conductance matrices for the corresponding linear elements, whereas

$$\widetilde{C}: \mathbb{R}^{n_{\widetilde{C}}} \to \mathbb{R}^{n_{\widetilde{C}},n_{\widetilde{C}}}, \qquad \widetilde{L}: \mathbb{R}^{n_{\widetilde{L}}} \to \mathbb{R}^{n_{\widetilde{L}},n_{\widetilde{L}}}, \qquad \widetilde{g}: \mathbb{R}^{n_{\widetilde{\mathcal{R}}}} \to \mathbb{R}^{n_{\widetilde{\mathcal{R}}}}$$

describe the corresponding nonlinear components, and $j_{\widetilde{L}}$ is the current vector through the nonlinear inductors.

2.2.3   Linear RLC Circuits

For simplification of notation, a linear RLC circuit containing n_R linear resistors, n_L linear inductors, n_C linear capacitors, n_I current sources and n_V voltage sources will be described by the linear DAE system

$$\begin{aligned} E\tfrac{d}{dt}x &= Ax + Bu, \\ y &= B^Tx, \end{aligned} \tag{2.9}$$

with the MNA matrices

$$E = \begin{bmatrix} A_C C A_C^T & 0 & 0 \\ 0 & L & 0 \\ 0 & 0 & 0 \end{bmatrix}, \quad A = \begin{bmatrix} -A_R G A_R^T & -A_L & -A_V \\ A_L^T & 0 & 0 \\ A_V^T & 0 & 0 \end{bmatrix}, \quad B = \begin{bmatrix} -A_I & 0 \\ 0 & 0 \\ 0 & -I \end{bmatrix}, \tag{2.10}$$

or the MLA matrices

$$E = \begin{bmatrix} B_L L B_L^T & 0 & 0 \\ 0 & C & 0 \\ 0 & 0 & 0 \end{bmatrix}, \quad A = \begin{bmatrix} -B_R R B_R^T & -B_C & -B_I \\ B_C^T & 0 & 0 \\ B_I^T & 0 & 0 \end{bmatrix}, \quad B = \begin{bmatrix} 0 & -B_V \\ 0 & 0 \\ -I & 0 \end{bmatrix}. \tag{2.11}$$

Here, the subscripts R, C, L, V and I stand for linear resistors, linear capacitors, linear inductors, voltage sources and current sources, respectively, and $L \in \mathbb{R}^{n_L,n_L}$, $C \in \mathbb{R}^{n_C,n_C}$, $R \in \mathbb{R}^{n_R,n_R}$ and $G = R^{-1}$ are the inductance, capacitance, resistance and conductance matrices, respectively. Linear circuits are often used to model interconnects, transmission lines and pin packages in VLSI networks. They arise also in the linearization of nonlinear circuit equations around DC operating points. According to (A1), we assume that

(A1′) the matrices L, C and G are symmetric and positive definite.

This condition together with (A2) and (A3) guarantees that the pencil $\lambda E - A$ is *regular*, i.e., $\det(\lambda E - A) \neq 0$ for some $\lambda \in \mathbb{C}$, see [27]. In this case, we can define a *transfer function*

$$\mathbf{G}(s) = B^T(sE - A)^{-1}B$$

of the DAE system (2.9). Applying the Laplace transform to (2.9) with an initial condition $x(0) = x_0$ satisfying $Ex_0 = 0$, we obtain $\mathbf{y}(s) = \mathbf{G}(s)\mathbf{u}(s)$, where $\mathbf{u}(s)$ and $\mathbf{y}(s)$ are the Laplace transformations of the input $u(t)$ and the output $y(t)$, respectively. Thus, the transfer function $\mathbf{G}(s)$ describes the input-output relation of (2.9) in the frequency domain. Note that the MNA system (2.9), (2.10) and the MLA system (2.9), (2.11) have the same transfer function.

For any rational matrix-valued function $\mathbf{G}(s)$, there exist matrices E, A, B_{in} and B_{out} such that $\mathbf{G}(s) = B_{out}(sE - A)^{-1}B_{in}$, see [21]. Then the DAE system

$$E\frac{\mathrm{d}}{\mathrm{d}t}x = Ax + B_{in}u,$$
$$y = B_{out}x,$$

is said to form a *realization* of $\mathbf{G}(s)$. We will also denote a realization of $\mathbf{G}(s)$ by $\mathbf{G} = (E, A, B_{in}, B_{out})$.

The transfer function $\mathbf{G}(s)$ is called *proper* if $\lim_{s \to \infty} \mathbf{G}(s) < \infty$, and *improper*, otherwise. If $\mathbf{G}(s)$ is proper and analytic in $\mathbb{C}_+$, then the $\mathscr{H}_\infty$-norm of $\mathbf{G}(s)$ is defined as

$$\|\mathbf{G}\|_{\mathscr{H}_\infty} = \sup_{s \in \mathbb{C}_+} \|\mathbf{G}(s)\| = \lim_{\substack{\sigma \to 0 \\ \sigma > 0}} \sup_{\omega \in \mathbb{R}} \|\mathbf{G}(\sigma + i\omega)\|,$$

where $\|\cdot\|$ denotes the spectral matrix norm.

2.2.3.1 Passivity

Passivity is the most important property of circuit equations. System (2.9) with $x(0) = 0$ is *passive* if

$$\int_0^t u(\tau)^T y(\tau)\, d\tau \geq 0 \tag{2.12}$$

for all $t \geq 0$ and all admissible u such that $u^T y$ is locally integrable. Passive elements can store and dissipate energy, but they do not generate energy. Thus, capacitors, resistors and inductors are passive, while current and voltage sources are not.

It is well known in linear network theory [1] that the DAE system (2.9) is passive if and only if its transfer function $\mathbf{G}(s) = B^T(sE - A)^{-1}B$ is *positive real*, i.e., $\mathbf{G}$ is analytic in $\mathbb{C}_+$ and $\mathbf{G}(s) + \mathbf{G}(s)^* \geq 0$ for all $s \in \mathbb{C}_+$. Since the system matrices in (2.10) satisfy $E = E^T \geq 0$ and $A + A^T \leq 0$, the transfer function of (2.9), (2.10) is positive real, and, hence, this system is passive.

2.2.3.2   Stability

Stability is a qualitative property of dynamical systems which describes the behaviour of their solutions under small perturbations in the initial data. For the linear DAE system (2.9), stability can be characterized in terms of the finite eigenvalues of the pencil $\lambda E - A$, e.g., [21]. System (2.9) is *stable* if all the finite eigenvalues of $\lambda E - A$ lie in the closed left half-plane and the eigenvalues on the imaginary axis are semi-simple, i.e., they have the same algebraic and geometric multiplicity. System (2.9) is *asymptotically stable* if the pencil $\lambda E - A$ is *c-stable*, i.e., all its finite eigenvalues lie in the open left half-plane. Note that passivity of the MNA system (2.9), (2.10) implies that this system is stable [1, Theorem 2.7.2]. Topological conditions for the asymptotic stability of the MNA equations (2.9), (2.10) can be found in [58, 59].

2.2.3.3   Reciprocity

Another relevant property of circuit equations is reciprocity. We call a matrix $S \in \mathbb{R}^{m,m}$ a *signature* if S is diagonal and $S^2 = I_m$. System (2.9) is *reciprocal* with an external signature $S_{\text{ext}} \in \mathbb{R}^{m,m}$ if its transfer function satisfies

$$\mathbf{G}(s) = S_{\text{ext}}\mathbf{G}(s)^T S_{\text{ext}}$$

for all $s \in \mathbb{C}$. Obviously, the MNA system (2.9), (2.10) with symmetric L, C and G is reciprocal with the external signature $S_{\text{ext}} = \text{diag}(I_{n_I}, -I_{n_V})$.

2.3   Model Reduction of Linear Circuits

Consider the linear MNA system (2.9), (2.10) with $E, A \in \mathbb{R}^{n,n}$ and $B \in \mathbb{R}^{n,m}$. We aim to approximate this system by a reduced-order model

$$\begin{aligned}
\hat{E}\tfrac{\mathrm{d}}{\mathrm{d}t}\hat{x} &= \hat{A}\hat{x} + \hat{B}u, \\
\hat{y} &= \hat{C}\hat{x},
\end{aligned} \qquad (2.13)$$

where $\hat{E}, \hat{A} \in \mathbb{R}^{n_r, n_r}$, $\hat{B} \in \mathbb{R}^{n_r, m}$, $\hat{C} \in \mathbb{R}^{m, n_r}$ and $n_r \ll n$. It is required that the approximate system (2.13) has a small approximation error $y - \hat{y}$ and also preserves passivity and reciprocity. In the frequency domain, the error can be measured via $\mathbf{G} - \hat{\mathbf{G}}$ in an appropriate system norm, where $\hat{\mathbf{G}}(s) = \hat{C}(s\hat{E} - \hat{A})^{-1}\hat{B}$ is the transfer function of system (2.13).

A classical approach for computing the reduced-order model (2.13) is based on the projection of system (2.9) onto lower dimensional subspaces. In this case, the system matrices in (2.13) have the form

$$\hat{E} = W^T E T, \qquad \hat{A} = W^T A T, \qquad \hat{B} = W^T B, \qquad \hat{C} = B^T T, \qquad (2.14)$$

where the projection matrices $W, T \in \mathbb{R}^{n, n_r}$ determine the subspaces of interest. In interpolation-based passivity-preserving model reduction methods like PRIMA [48], SPRIM [26, 27] and spectral zero interpolation [38, 66], the columns of these matrices span certain (rational) Krylov subspaces associated with (2.9).

Balanced truncation also belongs to the projection-based model reduction techniques. This method consists in transforming the dynamical system into a balanced form whose appropriately chosen controllability and observability Gramians are both equal to a diagonal matrix. Then a reduced-order model (2.13), (2.14) is obtained by projecting (2.9) onto the subspaces corresponding to the dominant diagonal elements of the balanced Gramians. In order to capture specific system properties, different balancing techniques have been developed in the last 30 years [31, 46, 47, 52, 55, 69]. An important property of these techniques is the existence of computable error bounds that allow us to approximate (2.9) to a prescribed accuracy.

In Sect. 2.3.1, we consider a passivity-preserving model reduction method for general RLC circuits developed in [54, 72]. This method is based on balancing the Gramians that satisfy the projected Lur'e matrix equations. For RC circuits consisting only of resistors, capacitors, current sources and/or voltage sources, this method can significantly be simplified. In Sect. 2.3.2, we present passivity-preserving model reduction methods for RC circuits developed in [56] that rely on balancing the solutions of the projected Lyapunov equations. Thereby, we will distinguish three cases: RC circuits with current sources (RCI circuits), RC circuits with voltage sources (RCV circuits) and RC circuits with both current and voltage sources (RCIV circuits). Finally, in Sect. 2.3.3, we discuss the numerical aspects of the presented balancing-related model reduction algorithms.

2.3.1 Balanced Truncation for RLC Circuits

First, we consider model reduction of general RLC circuits. Note that passivity of the MNA system (2.9), (2.10) can be characterized via the *projected Lur'e equations*

$$
\begin{aligned}
E X (A - BB^T)^T + (A - BB^T) XE^T + 2P_l BB^T P_l^T &= -2K_c K_c^T, \\
EXB - P_l BM_0^T = -K_c J_c^T, \quad I - M_0 M_0^T = J_c J_c^T, \quad X &= P_r XP_r^T \geq 0,
\end{aligned} \qquad (2.15)
$$

and

$$E^T Y (A - BB^T) + (A - BB^T)^T YE + 2P_r^T BB^T P_r = -2K_o^T K_o,$$
$$-E^T YB + P_r^T BM_0 = -K_o^T J_o, \quad I - M_0^T M_0 = J_o^T J_o \quad Y = P_l^T YP_l \geq 0 \tag{2.16}$$

with unknowns $X \in \mathbb{R}^{n,n}$, $K_c \in \mathbb{R}^{n,m}$, $J_c \in \mathbb{R}^{m,m}$ and $Y \in \mathbb{R}^{n,n}$, $K_o \in \mathbb{R}^{m,n}$, $J_o \in \mathbb{R}^{m,m}$, respectively. Here, P_r and P_l are the spectral projectors onto the right and left deflating subspaces of the pencil $\lambda E - (A - BB^T)$ corresponding to the finite eigenvalues along the right and left deflating subspaces corresponding to the eigenvalue at infinity, and

$$M_0 = I - 2 \lim_{s \to \infty} B^T (sE - A + BB^T)^{-1} B. \tag{2.17}$$

In general, the solvability of the projected Lur'e equations (2.15) and (2.16) requires that system (2.9) is passive and R-minimal, i.e.,

$$\mathrm{rank}([\lambda E - A, \ B]) = \mathrm{rank}([\lambda E^T - A^T, \ B]) = n$$

for all $\lambda \in \mathbb{C}$. For the circuit equations (2.9), (2.10), however, the R-minimality condition can be removed.

Theorem 2.3.1 ([54]) *Consider an MNA system (2.9), (2.10) satisfying (A1′), (A2) and (A3). Then the projected Lur'e equations (2.15) and (2.16) are solvable.*

Note that the solutions X and Y of (2.15) and (2.16) are not unique. However, there exist unique minimal solutions $X_{\min}$ and $Y_{\min}$ that satisfy $0 \leq X_{\min} \leq X$ and $0 \leq Y_{\min} \leq Y$ for all symmetric solutions X and Y of (2.15) and (2.16), respectively. These minimal solutions $X_{\min}$ and $Y_{\min}$ of (2.15) and (2.16), respectively, are called the *controllability* and *observability Gramians* of system (2.9). This system is called *balanced* if $X_{\min} = Y_{\min} = \mathrm{diag}(\Gamma, 0)$, where $\Gamma = \mathrm{diag}(\gamma_1, \ldots, \gamma_{n_f})$ with $\gamma_1 \geq \ldots \geq \gamma_{n_f} \geq 0$ and $n_f = \mathrm{rank}(P_r)$. The values γ_j are called the *characteristic values* of (2.9). Based on the energy interpretation of the Gramians $X_{\min}$ and $Y_{\min}$, see [55], one can conclude that the truncation of the states of a balanced system corresponding to the small characteristic values does not change the system properties significantly. The characteristic values and balancing transformation matrices can be determined from the singular value decomposition of the matrix $\tilde{Y}^T E \tilde{X}$, where $\tilde{X}$ and $\tilde{Y}$ are the Cholesky factors of the Gramians $X_{\min} = \tilde{X}\tilde{X}^T$ and $Y_{\min} = \tilde{Y}\tilde{Y}^T$. Taking into account the block structure of the MNA matrices in (2.10), we have $E^T = S_{\mathrm{int}} E S_{\mathrm{int}}$ and $A^T = S_{\mathrm{int}} A S_{\mathrm{int}}$ with

$$S_{\mathrm{int}} = \mathrm{diag}(I_{n_v - 1}, -I_{n_L}, -I_{n_V}). \tag{2.18}$$

This implies that $Y_{\min} = S_{\mathrm{int}} X_{\min} S_{\mathrm{int}}$. Then instead of the more expensive singular value decomposition of $\tilde{Y}^T E \tilde{X}$, we can compute the eigenvalue decomposition of the symmetric matrix $\tilde{X}^T S_{\mathrm{int}} E \tilde{X}$. In this case, the numerical solution of only one Lur'e equation is required. If λ_j are eigenvalues of $\tilde{X}^T S_{\mathrm{int}} E \tilde{X}$, then $\gamma_j = |\lambda_j|$. Thus, the reduced-order model (2.13), (2.14) can be determined by projecting (2.9) onto the subspaces corresponding to the dominant eigenvalues of $\tilde{X}^T S_{\mathrm{int}} E \tilde{X}$.

One can also truncate the states that are uncontrollable and unobservable at infinity. Such states do not contribute to the energy transfer from the input to the output, and, therefore, they can be removed from the system without changing the input-output relation [69, 71]. For general DAE systems, such states can be determined from the solution of certain projected discrete-time Lyapunov equations [69]. Exploiting again the structure of the MNA equations (2.9), (2.10), the required states can be determined from the eigenvalue decomposition of the symmetric matrix $(I - M_0)S_{\text{ext}}$ with

$$S_{\text{ext}} = \text{diag}(I_{n_I}, -I_{n_V}). \tag{2.19}$$

We summarize the resulting model reduction method for RLC circuits in Algorithm 2.1.

Algorithm 2.1 Passivity-preserving balanced truncation for RLC circuits

---

Given a passive MNA system (2.9) with E, A, B as in (2.10), compute a reduced-order model (2.13).

1. Compute the full-rank Cholesky factor $\tilde{X}$ of the minimal solution $X_{\min} = \tilde{X}\tilde{X}^T$ of the projected Lur'e equation (2.15).
2. Compute the eigenvalue decomposition

$$\tilde{X}^T S_{\text{int}} E \tilde{X} = [U_1, U_2] \begin{bmatrix} \Lambda_1 & 0 \\ 0 & \Lambda_2 \end{bmatrix} [U_1, U_2]^T,$$

   where S_{int} is as in (2.18), the matrix $[U_1, U_2]$ is orthogonal, $\Lambda_1 = \text{diag}(\lambda_1, \ldots, \lambda_r)$ and $\Lambda_2 = \text{diag}(\lambda_{r+1}, \ldots, \lambda_q)$.
3. Compute the eigenvalue decomposition

$$(I - M_0)S_{\text{ext}} = U_0 \Lambda_0 U_0^T,$$

   where M_0 is as in (2.17), S_{ext} is as in (2.19), U_0 is orthogonal and $\Lambda_0 = \text{diag}(\hat{\lambda}_1, \ldots, \hat{\lambda}_m)$.
4. Compute the reduced-order system (2.13) with

$$\hat{E} = \begin{bmatrix} I & 0 \\ 0 & 0 \end{bmatrix}, \qquad \hat{A} = \begin{bmatrix} W^T A T & W^T B C_\infty/\sqrt{2} \\ -B_\infty B^T T/\sqrt{2} & I - B_\infty C_\infty/2 \end{bmatrix},$$

$$\hat{B} = \begin{bmatrix} W^T B \\ -B_\infty/\sqrt{2} \end{bmatrix}, \qquad \hat{C} = \begin{bmatrix} B^T T, & C_\infty/\sqrt{2} \end{bmatrix},$$

   where

$$B_\infty = S_0 |\Lambda_0|^{1/2} U_0^T S_{\text{ext}}, \quad C_\infty = U_0 |\Lambda_0|^{1/2},$$

$$W = S_{\text{int}} \tilde{X} U_1 |\Lambda_1|^{-1/2}, \quad T = \tilde{X} U_1 S_1 |\Lambda_1|^{-1/2},$$

$$S_0 = \text{diag}(\text{sign}(\hat{\lambda}_1), \ldots, \text{sign}(\hat{\lambda}_m)), \quad |\Lambda_0| = \text{diag}(|\hat{\lambda}_1|, \ldots, |\hat{\lambda}_m|),$$

$$S_1 = \text{diag}(\text{sign}(\lambda_1), \ldots, \text{sign}(\lambda_r)), \quad |\Lambda_1| = \text{diag}(|\lambda_1|, \ldots, |\lambda_r|).$$

One can show that the reduced-order model computed by Algorithm 2.1 preserves not only passivity but also reciprocity. Moreover, we have the following error bound

$$\|\hat{\mathbf{G}} - \mathbf{G}\|_{\mathscr{H}_\infty} \leq \frac{\|I + \mathbf{G}\|^2_{\mathscr{H}_\infty}(\gamma_{r+1} + \ldots + \gamma_q)}{1 - \|I + \mathbf{G}\|_{\mathscr{H}_\infty}(\gamma_{r+1} + \ldots + \gamma_q)},$$

provided $\|I + \mathbf{G}\|_{\mathscr{H}_\infty}(\gamma_{r+1} + \ldots + \gamma_q) < 1$, see [54] for details. Note that this error bound requires the computation of the $\mathscr{H}_\infty$-norm of $\mathbf{G}$, which is expensive for large-scale systems. If r is chosen in Algorithm 2.1 such that

$$\|I + \hat{\mathbf{G}}\|_{\mathscr{H}_\infty}(\gamma_{r+1} + \ldots + \gamma_q) < 1,$$

then we can estimate

$$\|\hat{\mathbf{G}} - \mathbf{G}\|_{\mathscr{H}_\infty} \leq \frac{\|I + \hat{\mathbf{G}}\|^2_{\mathscr{H}_\infty}(\gamma_{r+1} + \ldots + \gamma_q)}{1 - \|I + \hat{\mathbf{G}}\|_{\mathscr{H}_\infty}(\gamma_{r+1} + \ldots + \gamma_q)}, \tag{2.20}$$

where only the evaluation of the $\mathscr{H}_\infty$-norm of the reduced-order system $\hat{\mathbf{G}}$ is required.

If the matrix $I - M_0 M_0^T$ is nonsingular, then the projected Lur'e equation (2.15) can be written as the *projected Riccati equation*

$$EXF^T + FXE^T + EXB_c^T B_c XE^T + P_l B_o B_o^T P_l^T = 0, \quad X = P_r X P_r^T, \tag{2.21}$$

where

$$\begin{aligned}
F &= A - BB^T - 2P_l BM_0^T (I - M_0 M_0^T)^{-1} B^T P_r, \\
B_c &= \sqrt{2} J_c^{-1} B^T P_r, \qquad B_o = -\sqrt{2} B J_o^{-1}, \\
J_c J_c^T &= I - M_0 M_0^T, \quad J_o^T J_o = I - M_0^T M_0.
\end{aligned} \tag{2.22}$$

Note that the invertibility of $I - M_0 M_0^T$ depends on the topological structure of the circuit.

Theorem 2.3.2 *Consider an MNA system* (2.9), (2.10). *Let the matrix M_0 be as in* (2.17). *Then $I - M_0 M_0^T$ is nonsingular if and only if*

$$\text{rank}(Z_C^T[A_I, A_V]) = n_I + n_V, \qquad Z_{RC}^T[A_I, A_V] = 0, \tag{2.23}$$

where Z_C and Z_{RC} are the basis matrices for $\ker(A_C^T)$ and $\ker([A_R, A_C]^T)$, respectively.

Proof The result immediately follows from [54, Theorem 7].

The first condition in (2.23) is equivalent to the absence of loops of capacitors, voltage sources and current sources (CVI-loops) except for loops consisting of capacitive branches (C-loops). The second condition in (2.23) means that the circuit does not contain cutsets consisting of branches of inductors, voltage sources and current sources (LVI-cutsets) except for cutsets consisting of inductive branches (L-cutsets).

2.3.2 Balanced Truncation for RC Circuits

We now present a Lyapunov-based balanced truncation model reduction approach for RC circuits. In this approach, the Gramians of system (2.9) are defined as unique symmetric, positive semidefinite solutions of the *projected continuous-time Lyapunov equations*

$$
EXA^T + AXE^T = -P_l BB^T P_l^T, \quad X = P_r X P_r^T,
$$
$$
E^T YA + A^T YE = -P_r^T BB^T P_r, \quad Y = P_l^T YP_l.
$$

The numerical solution of such equations is much less exhausting than of the projected Lur'e or Riccati equations. For a balanced system, these Gramians are both equal to a diagonal matrix

$$
X = Y = \mathrm{diag}(\Sigma, 0),
$$

where $\Sigma = \mathrm{diag}(\sigma_1, \ldots, \sigma_{n_f})$. The values σ_j are called the *proper Hankel singular values* of system $\mathbf{G} = (E, A, B, B^T)$. They determine which states are important and which states can be removed from the system.

Note that Lyapunov-based balanced truncation does not, in general, guarantee the preservation of passivity in the reduced-order model. However, the RC circuit equations either have a symmetric structure

$$
E = E^T \geq 0, \qquad A = A^T \leq 0, \tag{2.24}
$$

or they can be transformed under preservation of passivity into a symmetric form. Then Lyapunov-based balanced truncation applied to symmetric systems is known to be structure-preserving [45] and, hence, also passivity-preserving. All model reduction algorithms presented in this section have been developed in [56].

2.3.2.1 RCI Circuits

First, we consider RCI circuits consisting of resistors, capacitors and current sources only. The MNA matrices are then given by

$$
E = A_C C A_C^T, \qquad A = -A_R G A_R^T, \qquad B = -A_I. \tag{2.25}
$$

Algorithm 2.2 Passivity-preserving balanced truncation for RCI circuits

---

Given a passive MNA system (2.9) with E, A, B as in (2.25), compute a reduced-order model (2.13).

1. Compute the full column rank matrices Z_C and Z_R such that

$$\text{im}(Z_C) = \ker(A_C^T), \qquad \text{im}(Z_R) = \ker(A_R^T).$$

2. Compute a full-rank Cholesky factor $\tilde{X}$ of the solution $X = \tilde{X}\tilde{X}^T$ of the projected Lyapunov equation

$$EXA + EXA = -PBB^T P, \qquad X = PXP,$$

where

$$P = I - Z_R(Z_R^T E Z_R)^{-1} Z_R^T E - Z_C(Z_C^T A Z_C)^{-1} Z_C^T A$$

is the spectral projector onto the right deflating subspace of $\lambda E - A$ corresponding to the finite eigenvalues with negative real part.
3. Compute the eigenvalue decomposition

$$\tilde{X}^T E \tilde{X} = [U_1, U_2] \begin{bmatrix} \Sigma_1 & 0 \\ 0 & \Sigma_2 \end{bmatrix} [U_1, U_2]^T, \tag{2.26}$$

where $[U_1, U_2]$ is orthogonal, $\Sigma_1 = \text{diag}(\sigma_1, \ldots, \sigma_r)$ and $\Sigma_2 = \text{diag}(\sigma_{r+1}, \ldots, \sigma_q)$.
4. Compute the full-rank Cholesky factors $B_0 \in \mathbb{R}^{r_0, m}$ and $B_\infty \in \mathbb{R}^{r_\infty, m}$ of the matrices $R_0 = B_0^T B_0$ and $R_\infty = B_\infty^T B_\infty$ given by

$$R_0 = B^T Z_R(Z_R^T E Z_R)^{-1} Z_R^T B, \qquad R_\infty = -B^T Z_C(Z_C^T A Z_C)^{-1} Z_C^T B. \tag{2.27}$$

5. Compute the reduced-order system (2.13) with

$$\hat{E} = \begin{bmatrix} I_r & 0 & 0 \\ 0 & I_{r_0} & 0 \\ 0 & 0 & 0 \end{bmatrix}, \quad \hat{A} = \begin{bmatrix} A_s & 0 & 0 \\ 0 & 0 & 0 \\ 0 & 0 & -I_{r_\infty} \end{bmatrix}, \quad \hat{B} = \hat{C}^T = \begin{bmatrix} B_s \\ B_0 \\ B_\infty \end{bmatrix}, \tag{2.28}$$

where

$$A_s = \Sigma_1^{-1/2} U_1^T \tilde{X}^T A \tilde{X} U_1 \Sigma_1^{-1/2} \quad and \quad B_s = \Sigma_1^{-1/2} U_1^T \tilde{X}^T B. \tag{2.29}$$

Obviously, the symmetry condition (2.24) is fulfilled. In this case, the reduced-order system (2.13) can be computed by Algorithm 2.2.

One can show that the reduced-order system (2.13), (2.28) has the transfer function

$$\hat{\mathbf{G}}(s) = \hat{C}(s\hat{E} - \hat{A})^{-1}\hat{B} = B_s^T(sI - A_s)^{-1}B_s + \frac{1}{s}R_0 + R_\infty,$$

where the matrices R_0 and R_∞ are as in (2.27), and A_s and B_s are given in (2.29). Furthermore, we have the following $\mathcal{H}_\infty$-norm error bound.

Theorem 2.3.3 ([56]) *Let an RCI circuit* (2.9), (2.25) *fulfill* (A1′) *and* (A2). *Then a reduced-order model* (2.13), (2.28) *obtained by Algorithm 2.2 is passive and reciprocal with an external signature* $S_{\text{ext}} = I_{n_I}$. *Moreover, for the transfer functions* $\mathbf{G}$ *and* $\hat{\mathbf{G}}$ *of the original system* (2.9), (2.25) *and the reduced-order model* (2.13), (2.28), *we have the* $\mathcal{H}_\infty$-*norm error bound*

$$\|\mathbf{G} - \hat{\mathbf{G}}\|_{\mathcal{H}_\infty} \leq 2(\sigma_{r+1} + \ldots + \sigma_q),$$

where σ_j *are the proper Hankel singular values of* $\mathbf{G} = (E, A, B, B^T)$ *obtained in* (2.26).

2.3.2.2   RCV Circuits

We now consider RCV circuits consisting of resistors, capacitors and voltage sources. Unfortunately, the MNA equations for such circuits do not satisfy the symmetry conditions (2.24). We can, however, transform the MLA equations with the system matrices

$$E = \begin{bmatrix} 0 & 0 \\ 0 & C \end{bmatrix}, \qquad A = \begin{bmatrix} -B_R R B_R^T & -B_C \\ B_C^T & 0 \end{bmatrix}, \qquad B = \begin{bmatrix} -B_V \\ 0 \end{bmatrix} \qquad (2.30)$$

into a symmetric system. Such a transformation is the frequency inversion

$$\mathbf{G}^\star(s) = \mathbf{G}(s^{-1}).$$

The transfer function of the transformed system can be realized as

$$\mathbf{G}^\star(s) = B_\star^T (s E_\star - A_\star)^{-1} B_\star,$$

where

$$E_\star = B_C C^{-1} B_C^T, \qquad A_\star = -B_R R B_R^T, \qquad B_\star = -B_V. \qquad (2.31)$$

Reducing this system and applying the back transformation, we obtain a reduced-order model. The resulting model reduction method is given in Algorithm 2.3.

The following theorem provides the error bound for the reduced model (2.13), (2.33).

Algorithm 2.3 Passivity-preserving balanced truncation for RCV circuits

---

Given a passive MLA system (2.9) with E, A, B as in (2.30), compute a reduced-order model (2.13).

1. Compute the full column rank matrices Y_C and Y_R such that $\mathrm{im}(Y_C) = \ker(B_C^T)$ and $\mathrm{im}(Y_R) = \ker(B_R^T)$.
2. Compute a full-rank Cholesky factor $\tilde{X}$ of the solution $X = \tilde{X}\tilde{X}^T$ of the projected Lyapunov equation

$$E_\star X A_\star + E_\star X A_\star = -PB_\star B_\star^T P, \quad X = PXP,$$

   where $E_\star$, $A_\star$, $B_\star$ are as in (2.31) and

$$P = I - Y_R(Y_R^T E_\star Y_R)^{-1} Y_R^T E_\star - Y_C(Y_C^T A_\star Y_C)^{-1} Y_C^T A_\star.$$

   is the projector onto the right deflating subspace of $\lambda E_\star - A_\star$ corresponding to the finite eigenvalues with negative real part.
3. Compute the eigenvalue decomposition

$$\tilde{X}^T E_\star \tilde{X} = [U_1, U_2] \begin{bmatrix} \Sigma_1 & 0 \\ 0 & \Sigma_2 \end{bmatrix} [U_1, U_2]^T, \tag{2.32}$$

   where $[U_1, U_2]$ is orthogonal, $\Sigma_1 = \mathrm{diag}(\sigma_1, \dots, \sigma_r)$ and $\Sigma_2 = \mathrm{diag}(\sigma_{r+1}, \dots, \sigma_q)$.
4. Compute the matrices

$$A_s = \Sigma_1^{-1/2} U_1^T \tilde{X}^T A_\star \tilde{X} U_1 \Sigma_1^{-1/2}, \quad B_s = \Sigma_1^{-1/2} U_1^T \tilde{X}^T B_\star,$$
$$R_0 = B_\star^T Y_R (Y_R^T E_\star Y_R)^{-1} Y_R^T B_\star, \quad R_\infty = -B_\star^T Y_C (Y_C^T A_\star Y_C)^{-1} Y_C^T B_\star,$$
$$\tilde{R}_\infty = R_\infty - B_s^T A_s^{-1} B_s.$$

5. Compute the eigenvalue decomposition

$$\begin{bmatrix} \tilde{R}_\infty & R_0 \\ R_0 & 0 \end{bmatrix} = [V_1, V_2] \begin{bmatrix} \Lambda_0 & 0 \\ 0 & 0 \end{bmatrix} [V_1, V_2]^T,$$

   where $[V_1, V_2]$ is orthogonal and Λ_0 is nonsingular.
6. Compute the reduced-order system (2.13) with

$$\hat{E} = \begin{bmatrix} I & 0 \\ 0 & \hat{E}_\infty \end{bmatrix}, \quad \hat{A} = \begin{bmatrix} \hat{A}_1 & 0 \\ 0 & \hat{A}_\infty \end{bmatrix}, \quad \hat{B} = -\hat{C}^T = \begin{bmatrix} \hat{B}_1 \\ \hat{B}_\infty \end{bmatrix}, \tag{2.33}$$

   where $\hat{A}_1 = A_s^{-1}$, $\hat{B}_1 = A_s^{-1} B_s$ and

$$\hat{E}_\infty = V_1^T \begin{bmatrix} R_0 & 0 \\ 0 & 0 \end{bmatrix} V_1, \quad \hat{A}_\infty = V_1^T \begin{bmatrix} \tilde{R}_\infty & R_0 \\ R_0 & 0 \end{bmatrix} V_1, \quad \hat{B}_\infty = V_1^T \begin{bmatrix} \tilde{R}_\infty \\ R_0 \end{bmatrix}.$$

Theorem 2.3.4 ([56]) *Let an RCV circuit fulfill assumptions* (A1′) *and* (A3). *Then a reduced-order model* (2.13), (2.33) *obtained by Algorithm 2.3 is passive and reciprocal with an external signature* $S_{\text{ext}} = I_{n_V}$. *Moreover, for the transfer functions* **G** *and* **Ĝ** *of the original system* (2.9), (2.10) *and the reduced-order model* (2.13), (2.33), *we have the* $\mathcal{H}_\infty$-*norm error bound*

$$\|\mathbf{G} - \hat{\mathbf{G}}\|_{\mathcal{H}_\infty} \le 2(\sigma_{r+1} + \ldots + \sigma_q),$$

where σ_j *are the proper Hankel singular values of* $\mathbf{G}^\star = (E_\star, A_\star, B_\star, B_\star^T)$ *obtained in* (2.32).

2.3.2.3 RCIV Circuits

Finally, we consider RCIV circuits that contain resistors, capacitors and both current as well as voltage sources. Such circuits are modeled by the linear system (2.9) with the MNA matrices

$$E = \begin{bmatrix} A_C C A_C^T & 0 \\ 0 & 0 \end{bmatrix}, \quad A = \begin{bmatrix} -A_R G A_R^T & -A_V \\ A_V^T & 0 \end{bmatrix}, \quad B = \begin{bmatrix} -A_I & 0 \\ 0 & -I \end{bmatrix} \quad (2.34)$$

or the MLA matrices

$$E = \begin{bmatrix} 0 & 0 & 0 \\ 0 & C & 0 \\ 0 & 0 & 0 \end{bmatrix}, \quad A = \begin{bmatrix} -B_R R B_R^T & -B_C & -B_I \\ B_C^T & 0 & 0 \\ B_I^T & 0 & 0 \end{bmatrix}, \quad B = \begin{bmatrix} 0 & -B_V \\ 0 & 0 \\ -I & 0 \end{bmatrix}. \quad (2.35)$$

Due to the reciprocity, the transfer function of this system can be partitioned in blocks as

$$\mathbf{G}(s) = \begin{bmatrix} \mathbf{G}_{II}(s) & \mathbf{G}_{IV}(s) \\ -\mathbf{G}_{IV}^T(s) & \mathbf{G}_{VV}(s) \end{bmatrix},$$

see [56]. Assume that

(A4)   the circuit does not contain cutsets of current and voltage sources.

Then $\mathbf{G}_{VV}(s)$ is invertible and $\mathbf{G}(s)$ has a (2,2) *partial inverse* defined as

$$\mathbf{G}^{(2,2)}(s) = \begin{bmatrix} \mathbf{G}_{II}(s) + \mathbf{G}_{IV}(s)\,\mathbf{G}_{VV}^{-1}(s)\,\mathbf{G}_{IV}^T(s) & -\mathbf{G}_{IV}(s)\,\mathbf{G}_{VV}^{-1}(s) \\ -\mathbf{G}_{VV}^{-1}(s)\,\mathbf{G}_{IV}^T(s) & \mathbf{G}_{VV}^{-1}(s) \end{bmatrix}.$$

Algorithm 2.4 Passivity-preserving balanced truncation for RCIV circuits—I

Given a passive MNA system (2.9) with E, A, B as in (2.34), compute a reduced-order model (2.13).

1. Compute a reduced-order model $\hat{\mathbf{G}}^{(2,2)} = (\hat{E}_{2,2}, \hat{A}_{2,2}, \hat{B}_{2,2}, \hat{C}_{2,2})$ by applying Algorithm 2.2 to the system $\mathbf{G}^{(2,2)} = (E_{2,2}, A_{2,2}, B_{2,2}, B_{2,2}^T)$ as in (2.36).
2. Compute the reduced-order system (2.13) with

$$
\hat{E} = \begin{bmatrix} \hat{E}_{2,2} & 0 \\ 0 & 0 \end{bmatrix}, \qquad
\hat{A} = \begin{bmatrix} \hat{A}_{2,2} & \hat{B}_2 \\ -\hat{B}_2^T & 0 \end{bmatrix}, \qquad
\hat{B} = \hat{C}^T = \begin{bmatrix} \hat{B}_1 & 0 \\ 0 & -I_{nv} \end{bmatrix}. \qquad (2.37)
$$

where $\hat{B}_1 = \hat{B}_{2,2}[I_{n_l}, \ 0]^T$ and $\hat{B}_2 = \hat{B}_{2,2}[0, \ I_{nv}]^T$.

This rational function can be realized as $\mathbf{G}^{(2,2)}(s) = B_{2,2}^T (sE_{2,2} - A_{2,2})^{-1} B_{2,2}$ with

$$
E_{2,2} = A_C C A_C^T, \qquad A_{2,2} = -A_R G A_R^T, \qquad B_{2,2} = [-A_I, \ -A_V]. \qquad (2.36)
$$

Note that the (2,2) partial inversion can interpreted as the replacements of all voltage sources by current sources. The system $\mathbf{G}^{(2,2)} = (E_{2,2}, A_{2,2}, B_{2,2}, B_{2,2}^T)$ is symmetric and passive. Then applying balanced truncation to this system and reversing the voltage replacement, we obtain a required reduced-order model, see Algorithm 2.4.

The following theorem establishes the properties of the reduced-order model (2.13), (2.37) and gives an error bound.

Theorem 2.3.5 ([56]) *Consider an RCIV circuit fulfilling assumptions* (A1′), (A3) *and* (A4). *Let* Z_R *an* Z_C *be the basis matrices for* $\ker(A_R^T)$ *and* $\ker(A_C^T)$, *respectively, and let* Z_R' *be the basis matrix for* $\operatorname{im}(A_R)$. *Assume that* $A_I^T Z_R (Z_R^T A_C C A_C^T Z_R)^{-1} Z_R^T A_V = 0$ *and* $Z_C^T A_V$ *has full column rank. Then the reduced-order model* (2.13), (2.37) *obtained by Algorithm 2.4 is passive and reciprocal with the external signature* $S_{\text{ext}} = \operatorname{diag}(I_{n_l}, -I_{nv})$. *Moreover, for the transfer functions* $\mathbf{G}$ *and* $\hat{\mathbf{G}}$ *of the original system* (2.9), (2.34) *and the reduced-order model* (2.13), (2.37), *we have the error bound*

$$
\|\mathbf{G} - \hat{\mathbf{G}}\|_{\mathscr{H}_\infty} \le 2\left(1 + c_1^2 + c_1^2 c_2^2\right)(\sigma_{r+1} + \ldots + \sigma_q),
$$

where σ_j *are the proper Hankel singular values of the* (2,2) *partially inverted system* $\mathbf{G}^{(2,2)}$,

$$
c_1 = \|(A_V^T Z_C (Z_C^T A_R G A_R^T Z_C)^{-1} Z_C^T A_V)^{-1}\|,
$$

$$
c_2 = \|A_V^T H A_V\|^{1/2} \|A_I^T H A_I\|^{1/2}
$$

with

$$H = QZ_R' \left((Z_R')^T A_R G A_R^T Z_R'\right)^{-1} (Z_R')^T Q^T \text{ and } Q = I - Z_R(Z_R^T A_C C A_C^T Z_R)^{-1} Z_R^T A_C C A_C^T.$$

An alternative approach for model reduction of RCIV circuits is based on considering the frequency-inverted MLA system

$$\mathbf{G}^\star(s) = \mathbf{G}(s^{-1}) = B_\star^T(sE_\star - A_\star)^{-1} B_\star$$

with the matrices

$$E_\star = \begin{bmatrix} B_C C^{-1} B_C^T & 0 \\ 0 & 0 \end{bmatrix}, \quad A_\star = \begin{bmatrix} -B_R R B_R^T & -B_I \\ B_I^T & 0 \end{bmatrix}, \quad B_\star = \begin{bmatrix} 0 & -B_V \\ -I & 0 \end{bmatrix}.$$

Let $\mathbf{G}^\star(s)$ be partitioned in blocks as

$$\mathbf{G}^\star(s) = \begin{bmatrix} \mathbf{G}_{11}(s) & \mathbf{G}_{12}(s) \\ -\mathbf{G}_{12}^T(s) & \mathbf{G}_{22}(s) \end{bmatrix}.$$

Assume that

(A5) the circuit does not contain loops of current and voltage sources.

Then $\mathbf{G}_{11}(s)$ is invertible and $\mathbf{G}^\star(s)$ has an $(1,1)$ *partial inverse* defined as

$$(\mathbf{G}^\star)^{(1,1)}(s) = \begin{bmatrix} \mathbf{G}_{11}^{-1}(s) & \mathbf{G}_{11}^{-1}(s) \mathbf{G}_{12}(s) \\ \mathbf{G}_{12}^T(s) \mathbf{G}_{11}^{-1}(s) & \mathbf{G}_{22}(s) + \mathbf{G}_{12}^T(s) \mathbf{G}_{11}^{-1}(s) \mathbf{G}_{12}(s) \end{bmatrix}$$
$$= B_{1,1}^T (s E_{1,1} - A_{1,1})^{-1} B_{1,1},$$

where

$$E_{1,1} = B_C C^{-1} B_C^T, \qquad A_{1,1} = -B_R R B_R^T, \qquad B_{1,1} = [-B_I, -B_V]. \qquad (2.38)$$

Reducing this symmetric system and reversing the initial transformation, we obtain a required reduced-order model. This model reduction method is presented in Algorithm 2.5.

The following theorem establishes the properties of the reduced-order model (2.13), (2.39) and gives an error bound.

Theorem 2.3.6 ([56]) *Consider an RCIV circuit fulfilling assumptions* (A1'), (A2) *and* (A5). *Let* Y_R *and* Y_C *be the basis matrices for* $\ker(B_R^T)$ *and* $\ker(B_C^T)$, *respectively, and let* Y_R' *be the basis matrix for* $\mathrm{im}(B_R)$. *Assume that* $B_V^T Y_R (Y_R^T B_C C^{-1} B_C^T Y_R)^{-1} Y_R^T B_I = 0$ *and* $Y_C^T B_I$ *has full column rank. Then the reduced-order model* (2.13), (2.39) *obtained by Algorithm 2.5 is passive and*

Algorithm 2.5 Passivity-preserving balanced truncation for RCIV circuits—II

Given a passive MLA system (2.9) with E, A, B as in (2.35), compute a reduced-order model (2.13).

1. Compute a reduced-order model $\hat{\mathbf{G}}_1 = (\hat{E}_1, \hat{A}_1, \hat{B}_1, \hat{C}_1)$ using Algorithm 2.3, where $E_\star$, $A_\star$ and $B_\star$ are replaced, respectively, by $E_{1,1}, A_{1,1}$ and $B_{1,1}$ as in (2.38).
2. Compute the reduced-order system (2.13) with

$$
\hat{E} = \begin{bmatrix} \hat{E}_1 & 0 \\ 0 & 0 \end{bmatrix}, \quad \hat{A} = \begin{bmatrix} \hat{A}_1 & \hat{B}_{11} \\ \hat{C}_{11} & 0 \end{bmatrix}, \quad \hat{B} = \begin{bmatrix} 0 & \hat{B}_{12} \\ I_{n_I} & 0 \end{bmatrix}, \quad \hat{C} = \begin{bmatrix} 0 & -I_{n_I} \\ \hat{C}_{21} & 0 \end{bmatrix}, \tag{2.39}
$$

where $\hat{B}_{11} = \hat{B}_1[I_{n_I}, 0]^T$, $\hat{B}_{12} = \hat{B}_1[0, I_{n_V}]^T$, $\hat{C}_{11} = [I_{n_I}, 0]\hat{C}_1$ and $\hat{C}_{21} = [0, I_{n_V}]\hat{C}_1$.

---

reciprocal with the external signature $S_{\text{ext}} = \text{diag}(I_{n_I}, -I_{n_V})$. *Moreover, for the transfer functions* $\mathbf{G}$ *and* $\hat{\mathbf{G}}$ *of the original system* (2.9), (2.35) *and the reduced-order model* (2.13), (2.39), *we have the error bound*

$$
\|\mathbf{G} - \hat{\mathbf{G}}\|_{\mathscr{H}_\infty} \le 2(1 + \tilde{c}_1^2 + \tilde{c}_1^2 \tilde{c}_2^2)(\sigma_{r+1} + \ldots + \sigma_q),
$$

where σ_j *are the proper Hankel singular values of the system* $(\mathbf{G}^\star)^{(1,1)}$,

$$
\tilde{c}_1 = \|(B_I^T Y_C (Y_C^T B_R R B_R^T Y_C)^{-1} Y_C^T B_I)^{-1}\|,
$$

$$
\tilde{c}_2 = \|B_V^T \tilde{H} B_V\|^{1/2} \|B_I^T \tilde{H} B_I\|^{1/2}
$$

with

$$
\tilde{H} = \tilde{Q} Y_R' \left((Y_R')^T B_R R B_R^T Y_R'\right)^{-1} (Y_R')^T \tilde{Q}^T, \tilde{Q} = I - Y_R (Y_R^T B_C C^{-1} B_C^T Y_R)^{-1} Y_R^T B_C C^{-1} B_C^T.
$$

Remark 2.3.7 Model reduction methods for RC circuits can also be extended to RL circuits which contain resistors, inductors, voltage and/or current sources. Observing that the frequency-inverted MNA equations for RLI circuits as well as the MLA equations for RLV circuits yield symmetric systems, we can design balanced truncation model reduction methods for RL circuits similar to Algorithms 2.2–2.5.

2.3.3 Numerical Aspects

The most expensive step in the presented model reduction algorithms is solving matrix equations. The numerical solution of the projected Lyapunov and Riccati equations will be discussed in Sects. 2.5.1 and 2.5.2, respectively. Here, we consider the computation of the matrix M_0 and the projectors P_r and P_l required in Algorithm 2.1 as well as the basis matrices required in Algorithms 2.2–2.5.

Fortunately, using the MNA structure of the system matrices in (2.10), the matrix M_0 and the projectors P_r and P_l can be computed in explicit form

$$M_0 = \begin{bmatrix} I - 2A_I^T Z H_0^{-1} Z^T A_I & 2A_I^T Z H_0^{-1} Z^T A_V \\ -2A_V^T Z H_0^{-1} Z^T A_I & -I + 2A_V^T Z H_0^{-1} Z^T A_V \end{bmatrix}, \tag{2.40}$$

$$P_r = \begin{bmatrix} H_5(H_4 H_2 - I) & H_5 H_4 A_L H_6 & 0 \\ 0 & H_6 & 0 \\ -A_V^T(H_4 H_2 - I) & -A_V^T H_4 A_L H_6 & 0 \end{bmatrix} = S_{\text{int}} P_l^T S_{\text{int}}, \tag{2.41}$$

where S_{int} is given in (2.18), and

$$H_0 = Z^T (A_R G A_R^T + A_I A_I^T + A_V A_V^T) Z,$$
$$H_1 = Z_{CRIV}^T A_L L^{-1} A_L^T Z_{CRIV},$$
$$H_2 = A_R G A_R^T + A_I A_I^T + A_V A_V^T + A_L L^{-1} A_L^T Z_{CRIV} H_1^{-1} Z_{CRIV}^T A_L L^{-1} A_L^T,$$
$$H_3 = Z_C^T H_2 Z_C,$$
$$H_4 = Z_C H_3^{-1} Z_C^T,$$
$$H_5 = Z_{CRIV} H_1^{-1} Z_{CRIV}^T A_L L^{-1} A_L^T - I,$$
$$H_6 = I - L^{-1} A_L^T Z_{CRIV} H_1^{-1} Z_{CRIV}^T A_L,$$
$$Z = Z_C Z'_{RIV-C},$$

Z_C is a basis matrix for $\ker(A_C^T)$,

Z'_{RIV-C} is a basis matrix for $\mathrm{im}(Z_C^T[A_R, A_I, A_V])$,

Z_{CRIV} is a basis matrix for $\ker([A_C, A_R, A_I, A_V]^T)$,

see [54, 72] for details. The basis matrices Z_C and Z_{CRIV} can be computed by analyzing the corresponding subgraphs of the given network graph as described in [23]. For example, the matrix Z_C can be constructed in the form

$$Z_C = \Pi_C \begin{bmatrix} \mathbf{1}_{k_1} & & \\ & \ddots & \\ & & \mathbf{1}_{k_s} \\ & 0 & \end{bmatrix}$$

by searching the components of connectivity in the C-subgraph consisting of the capacitive branches only. Here, $\mathbf{1}_{k_i} = [1, \ldots, 1]^T \in \mathbb{R}^{k_i}$, $i = 1, \ldots, s$, and Π_C is a permutation matrix. For this purpose, we can use graph search algorithms like breadth-first-search [40]. As a consequence, the nonzero columns of

$$A_{RIV-C} = Z_C^T[A_R, A_I, A_V]$$

form again an incidence matrix. In order to compute the basis matrix Z'_{RIV-C}, we first determine the basis matrix

$$Z_{RIV-C} = \Pi_{RIV-C} \begin{bmatrix} \mathbf{1}_{l_1} & & & \\ & \ddots & & \\ & & \mathbf{1}_{l_t} & \\ & 0 & & \end{bmatrix}$$

for $\ker(A^T_{RIV-C})$ from the associated graph. Then the complementary matrix Z'_{RIV-C} can be determined as

$$Z'_{RIV-C} = \Pi_{RIV-C} S_{RIV-C},$$

where S_{RIV-C} is a selector matrix constructed from the identity matrix by removing 1-st, (l_1+1)-st, $\ldots$, $(l_1+\ldots+l_t+1)$-st columns. One can see that the resulting basis matrices and also the matrices H_0, H_1, H_2, H_3, H_5 and H_6 are sparse. Of course, the projector P_r will never be constructed explicitly. Instead, we use projector-vector products required in the numerical solution of the Riccati equation.

Algorithms 2.3 and 2.5 require the knowledge of the reduced loop matrix $\mathbf{B}$ that can be obtained by the search for a loop basis in the circuit graph [2, 22, 39, 40]. Since the efficiency in the numerical solution of the projected Lyapunov equations can be improved if the matrix coefficients are sparse, it is preferable to choose a basis of loops with length as small as possible. This kind of problem was treated in [49].

The basis matrices Z_R, Z'_R and Z_C required in Algorithms 2.2 and 2.4 can be computed using graph search algorithms as described above. The basis matrices Y_R and Y_C required in Algorithms 2.3 and 2.5 can be determined by searching for dependent loops in the graphs $\mathfrak{G}_R$ and $\mathfrak{G}_C$ consisting of the resistive and capacitive branches, respectively. Furthermore, the basis matrix Y'_R can be obtained by removing the linear dependent columns of B_R. Such columns can be determined by searching for cutsets in the graph $\mathfrak{G}_R$, e.g., [2]. For the analysis of loop dependency and the search for cutsets in a graph, there exist a variety of efficient algorithms, see [40] and the references therein.

2.4   Model Reduction of Nonlinear Circuits

In this section, we present a model reduction approach for nonlinear circuits containing large linear subnetworks. This approach is based on decoupling the nonlinear circuit equations (2.2) into linear and nonlinear subsystems in an appropriate way. The linear part is then approximated by a reduced-order model using one of the model reduction algorithms from Sect. 2.3 depending on the topological structure of the linear subcircuit. The nonlinear part either remains unchanged or is approximated by a trajectory piece-wise linear (TPWL) approach [57] based on

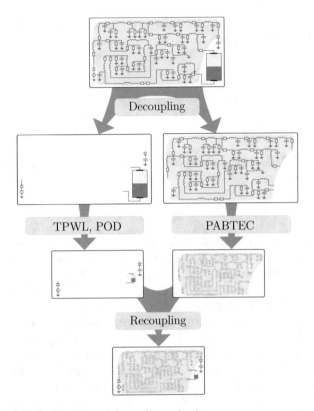

Fig. 2.1 A model reduction approach for nonlinear circuits

linearization, or proper orthogonal decomposition (POD), e.g., [65], which relies
on snapshot calculations. If the circuit contains semiconductor devices modeled by
instationary nonlinear partial differential equations [67, 73], these equations can
first be discretized in space and then reduced using the POD method as described in
[35, 36], see also Chap. 1 in this book. Finally, combining the reduced-order linear
and nonlinear models, we obtain a reduced-order nonlinear model that approximates
the MNA system (2.2). The concept of this model reduction approach is illustrated
in Fig. 2.1. We now describe this approach in more detail.

First, we consider the decoupling procedure developed in [68] that allows us
to extract a linear subcircuit from a nonlinear circuit. This procedure is based
on the formal replacement of nonlinear inductors by controlled current sources,
nonlinear capacitors by controlled voltage sources and nonlinear resistors by
equivalent circuits consisting of two serial linear resistors and one controlled
current source connected parallel to one of the introduced linear resistors. Such
replacements are demonstrated in Fig. 2.2, where we present two circuits before and
after replacements. It should be noted that the suggested replacements introduce
additional nodes and state variables, but neither additional CV-loops nor LI-cutsets

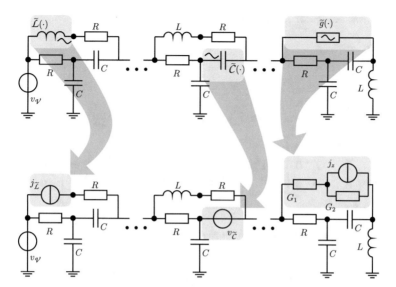

Fig. 2.2 Replacements of nonlinear circuit elements

occur in the decoupled linear subcircuit meaning that its index does not exceed the index of the original system (2.2). The following theorem establishes the decoupling on the equation level.

Theorem 2.4.1 [68] *Let $A^1_{\widetilde{\mathcal{R}}} \in \{0, 1\}^{n_v - 1, n_{\widetilde{\mathcal{R}}}}$ and $A^2_{\widetilde{\mathcal{R}}} \in \{-1, 0\}^{n_v - 1, n_{\widetilde{\mathcal{R}}}}$ satisfy the relation $A^1_{\widetilde{\mathcal{R}}} + A^2_{\widetilde{\mathcal{R}}} = A_{\widetilde{\mathcal{R}}}$, and let $G_1, G_2 \in \mathbb{R}^{n_{\widetilde{\mathcal{R}}}, n_{\widetilde{\mathcal{R}}}}$ be symmetric, positive definite. Assume that $v_{\widetilde{C}} \in \mathbb{R}^{n_{\widetilde{C}}}$ and $j_z \in \mathbb{R}^{n_{\widetilde{\mathcal{R}}}}$ satisfy*

$$v_{\widetilde{C}} = A^T_{\widetilde{C}} \eta,$$

$$j_z = (G_1 + G_2) G_1^{-1} \widetilde{g}(A^T_{\widetilde{\mathcal{R}}} \eta) - G_2 A^T_{\widetilde{\mathcal{R}}} \eta. \tag{2.42}$$

Then system (2.2) together with the relations

$$j_{\widetilde{C}} = \widetilde{C}(v_{\widetilde{C}}) \tfrac{\mathrm{d}}{\mathrm{d}t} v_{\widetilde{C}}, \tag{2.43}$$

$$\eta_z = (G_1 + G_2)^{-1} (G_1 (A^1_{\widetilde{\mathcal{R}}})^T \eta - G_2 (A^2_{\widetilde{\mathcal{R}}})^T \eta - j_z) \tag{2.44}$$

for the additional unknowns $\eta_z \in \mathbb{R}^{n_{\widetilde{\mathcal{R}}}}$ and $j_{\widetilde{C}} \in \mathbb{R}^{n_{\widetilde{C}}}$ has the same components η, η_z, $j_{\widetilde{L}}, j_{\widetilde{L}}, j_V$ and $j_{\widetilde{C}}$ in the state vector as the system

$$\widetilde{L}(j_{\widetilde{L}}) \tfrac{\mathrm{d}}{\mathrm{d}t} j_{\widetilde{L}} = A^T_{\widetilde{L}} \eta \tag{2.45}$$

coupled with the linear DAE system

$$E\tfrac{d}{dt}x_\ell = Ax_\ell + Bu_\ell,$$
$$y_\ell = B^T x_\ell, \tag{2.46}$$

where $x_\ell = \left[\eta^T \; \eta_z^T \middle| j_{\tilde{L}}^T \middle| j_{\mathcal{V}}^T \; j_{\tilde{C}}^T \right]^T$, $u_\ell = \left[j_{\mathcal{I}}^T \; j_z^T \; j_{\tilde{L}}^T \middle| v_{\mathcal{V}}^T \; v_{\tilde{C}}^T \right]^T$,

$$E = \begin{bmatrix} A_C C A_C^T & 0 & 0 \\ 0 & L & 0 \\ 0 & 0 & 0 \end{bmatrix}, \quad A = \begin{bmatrix} -A_R G A_R^T & -A_L & -A_V \\ A_L^T & 0 & 0 \\ A_V^T & 0 & 0 \end{bmatrix}, \quad B = \begin{bmatrix} -A_I & 0 \\ 0 & 0 \\ 0 & -I \end{bmatrix}, \tag{2.47}$$

and the incidence and element matrices are given by

$$A_C = \begin{bmatrix} A_{\tilde{C}} \\ 0 \end{bmatrix}, \quad A_R = \begin{bmatrix} A_{\tilde{R}} & A_{\mathcal{R}}^1 & A_{\mathcal{R}}^2 \\ 0 & -I & I \end{bmatrix}, \quad A_L = \begin{bmatrix} A_{\tilde{L}} \\ 0 \end{bmatrix}, \quad A_V = \begin{bmatrix} A_{\mathcal{V}} & A_{\tilde{C}} \\ 0 & 0 \end{bmatrix},$$

$$A_I = \begin{bmatrix} A_{\mathcal{J}} & A_{\mathcal{R}}^2 & A_{\tilde{L}} \\ 0 & I & 0 \end{bmatrix}, \quad G = \begin{bmatrix} \bar{G} & 0 & 0 \\ 0 & G_1 & 0 \\ 0 & 0 & G_2 \end{bmatrix}, \quad C = \bar{C}, \quad L = \bar{L}. \tag{2.48}$$

Note that the system matrices in the decoupled linear system (2.46)–(2.48) are in the MNA form. This system has the state space dimension

$$n_\ell = (n_v - 1) + n_{\tilde{\mathcal{R}}} + n_{\tilde{L}} + n_{\mathcal{V}} + n_{\tilde{C}}$$

and the input space dimension $m_\ell = n_{\mathcal{J}} + n_{\tilde{\mathcal{R}}} + n_{\tilde{L}} + n_{\mathcal{V}} + n_{\tilde{C}}$. It should also be noted that the state equivalence in Theorem 2.4.1 is independent of the choice of the matrices G_1 and G_2 satisfying the assumptions in the theorem. The substitution of nonlinear resistors with equivalent circuits as described above implies that G_1 and G_2 are both diagonal and their diagonal elements are conductances of the first and the second linear resistors, respectively, in the replacement circuits.

The following theorem establishes the well-posedness of the decoupled system (2.46)–(2.48).

Theorem 2.4.2 ([68]) *Let a nonlinear circuit satisfy assumptions (A1)–(A3). Assume that it contains neither loops consisting of nonlinear capacitors and voltage sources ($\widetilde{C}V$-loops) nor cutsets of nonlinear inductors and/or current sources ($\widetilde{C}V$-loops). Then the decoupled linear DAE system (2.46)–(2.48) modeling the linear subcircuit is well-posed in the sense that*

1. *the matrices C, L and G are symmetric and positive definite,*
2. *the matrix A_V has full column rank,*

3. the matrix $[A_C, A_L, A_R, A_V]$ *has full row rank.*

Note that the presence of $\widetilde{C}$V-loops and $\widetilde{L}$I-cutsets in the original circuit would lead after the replacement of the nonlinear capacitors and nonlinear inductors by voltage sources and current sources, respectively, to V-loops and I-cutsets in the decoupled circuit that would violate its well-posedness.

The next theorem shows that slightly stronger conditions for the original nonlinear circuit guarantee that the decoupled linear DAE system (2.46)–(2.48) is well-posed and, in addition, has index at most one.

Theorem 2.4.3 ([68]) *Let a nonlinear circuit satisfy assumptions* (A1)–(A3). *If this circuit contains neither* CV-*loops except for* $\bar{C}$-*loops with linear capacitors nor* LI-*cutsets, then the linear system* (2.46)–(2.48) *modeling the extracted linear subcircuit is well-posed and is of index at most one.*

The index one condition for system (2.46)–(2.48) implies that its transfer function is proper. The approximation of such systems is much easier than that of systems with an improper transfer function [71].

Depending on the topology of the extracted linear subcircuit, we can now apply one of the model reduction algorithms presented in Sect. 2.3 to the linear system (2.46)–(2.48). As a result, we obtain a reduced-order model (2.13) which can be combined with the nonlinear subsystem in order to get a reduced-order nonlinear model.

According to the block structure of the input and output vectors of the extracted linear DAE system (2.46)–(2.48), the reduced-order model (2.13) can be written in the form

$$\hat{E}\frac{\mathrm{d}}{\mathrm{d}t}\hat{x}_\ell = \hat{A}\hat{x}_\ell + [\hat{B}_1, \hat{B}_2, \hat{B}_3, \hat{B}_4, \hat{B}_5] \begin{bmatrix} j_{\mathcal{I}} \\ j_z \\ j_{\tilde{L}} \\ v_{\mathcal{V}} \\ v_{\tilde{C}} \end{bmatrix}, \qquad \begin{bmatrix} \hat{y}_{\ell 1} \\ \hat{y}_{\ell 2} \\ \hat{y}_{\ell 3} \\ \hat{y}_{\ell 4} \\ \hat{y}_{\ell 5} \end{bmatrix} = \begin{bmatrix} \hat{C}_1 \\ \hat{C}_2 \\ \hat{C}_3 \\ \hat{C}_4 \\ \hat{C}_5 \end{bmatrix} \hat{x}_\ell,$$

$$(2.49)$$

where $\hat{y}_{\ell j} = \hat{C}_j \hat{x}_\ell$, $j = 1, \ldots, 5$, approximate the corresponding components of the output of (2.46). Taking into account that $-A_{\tilde{L}}^T \eta \approx \hat{C}_3 \hat{x}_\ell$ and $-j_{\tilde{C}} \approx \hat{C}_5 \hat{x}_\ell$, Eqs. (2.43) and (2.45) are approximated by

$$\widetilde{C}(\hat{v}_{\tilde{C}})\frac{\mathrm{d}}{\mathrm{d}t}\hat{v}_{\tilde{C}} = -\hat{C}_5\hat{x}_\ell, \qquad \widetilde{L}(\hat{j}_{\tilde{L}})\frac{\mathrm{d}}{\mathrm{d}t}\hat{j}_{\tilde{L}} = -\hat{C}_3\hat{x}_\ell, \qquad (2.50)$$

respectively, where $\hat{j}_{\tilde{L}}$ and $\hat{v}_{\tilde{C}}$ are approximations to $j_{\tilde{L}}$ and $v_{\tilde{C}}$, respectively. Furthermore, for j_z and η_z defined in (2.42) and (2.44), respectively, we have

$$-(A_{\tilde{\mathcal{R}}}^2)^T\eta - \eta_z = -A_{\tilde{\mathcal{R}}}^T\eta + G_1^{-1}\widetilde{g}(A_{\tilde{\mathcal{R}}}^T\eta) = -v_{\tilde{\mathcal{R}}} + G_1^{-1}\widetilde{g}(v_{\tilde{\mathcal{R}}}).$$

Since $-(A^2_{\widetilde{\mathcal{R}}})^T\eta - \eta_z \approx \hat{C}_2\hat{x}_\ell$, this equation is approximated by

$$0 = -G_1\hat{C}_2\hat{x}_\ell - G_1\hat{v}_{\widetilde{\mathcal{R}}} + \widetilde{g}(\hat{v}_{\widetilde{\mathcal{R}}}), \tag{2.51}$$

where $\hat{v}_{\widetilde{\mathcal{R}}}$ approximates $v_{\widetilde{\mathcal{R}}}$. Combining (2.45), (2.49), (2.50) and (2.51), we obtain the reduced-order nonlinear model

$$\begin{aligned}\hat{\mathscr{E}}(\hat{x})\tfrac{d}{dt}\hat{x} &= \hat{\mathscr{A}}\hat{x} + \hat{f}(\hat{x}) + \hat{\mathscr{B}}u,\\ \hat{y} &= \hat{\mathscr{C}}\hat{x},\end{aligned} \tag{2.52}$$

where $\hat{x} = \left[\begin{array}{cccc}\hat{x}_\ell^T, & \hat{j}_L^T, & \hat{v}_C^T, & \hat{v}_{\widetilde{\mathcal{R}}}^T\end{array}\right]^T$, $u = \left[\begin{array}{cc}j_j^T, & v_\mathcal{V}^T\end{array}\right]^T$ and

$$\hat{\mathscr{E}}(\hat{x}) = \begin{bmatrix}\hat{E} & 0 & 0 & 0\\ 0 & \widetilde{L}(\hat{j}_{\widetilde{L}}) & 0 & 0\\ 0 & 0 & \widetilde{C}(\hat{v}_{\widetilde{C}}) & 0\\ 0 & 0 & 0 & 0\end{bmatrix}, \quad \hat{f}(\hat{x}) = \begin{bmatrix}0\\ 0\\ 0\\ \widetilde{g}(\hat{v}_{\widetilde{\mathcal{R}}})\end{bmatrix}, \quad \hat{\mathscr{B}} = \begin{bmatrix}\hat{B}_1 & \hat{B}_4\\ 0 & 0\\ 0 & 0\\ 0 & 0\end{bmatrix}, \tag{2.53}$$

$$\hat{\mathscr{A}} = \begin{bmatrix}\hat{A} + \hat{B}_2(G_1 + G_2)\hat{C}_2 & \hat{B}_3 & \hat{B}_5 & \hat{B}_2G_1\\ -\hat{C}_3 & 0 & 0 & 0\\ -\hat{C}_5 & 0 & 0 & 0\\ -G_1\hat{C}_2 & 0 & 0 & -G_1\end{bmatrix}, \quad \hat{\mathscr{C}} = \begin{bmatrix}\hat{C}_1 & 0 & 0 & 0\\ \hat{C}_4 & 0 & 0 & 0\end{bmatrix}. \tag{2.54}$$

This model can now be used for further investigations in steady-state analysis, transient analysis or sensitivity analysis of electrical circuits. Note that the error bounds for the reduced-order linear subsystem (2.49) presented in Sect. 2.3 can be used to estimate the error in the output of the reduced-order nonlinear system (2.52)–(2.54), see [34] for such estimates for a special class of nonlinear circuits.

2.5   Solving Matrix Equations

In this section, we consider numerical algorithms for solving the projected Lyapunov and Riccati equations developed in [55, 70, 71]. In practice, the numerical rank of the solutions of these equations is often much smaller than the dimension of the problem. Then such solutions can be well approximated by low-rank matrices. Moreover, these low-rank approximations can be determined directly in factored form. Replacing the Cholesky factors of the Gramians in Algorithms 2.1–2.5 by their low-rank factors reduces significantly the computational complexity and storage requirements in the balancing-related model reduction methods and makes these methods very suitable for large-scale circuit equations.

2.5.1 ADI Method for Projected Lyapunov Equations

We focus first on solving the projected Lyapunov equation

$$EXA^T + AXE^T = -P_lBB^TP_l^T, \qquad X = P_rXP_r^T \qquad (2.55)$$

using the *alternating direction implicit* (ADI) *method*. Such an equation has to be solved in Algorithms 2.2–2.5. The ADI method has been first proposed for standard Lyapunov equations [14, 44, 50, 76] and then extended in [70] to projected Lyapunov equations. The generalized ADI iteration for the projected Lyapunov equation (2.55) is given by

$$
\begin{aligned}
(E + \tau_k A)X_{k-1/2}A^T + AX_{k-1}(E - \tau_k A)^T &= -P_lBB^TP_l^T, \\
(E + \overline{\tau}_k A)X_k^T A^T + AX_{k-1/2}^T(E - \overline{\tau}_k A)^T &= -P_lBB^TP_l^T
\end{aligned}
\qquad (2.56)
$$

with an initial matrix $X_0 = 0$ and shift parameters $\tau_1, \ldots, \tau_k \in \mathbb{C}_-$. Here, $\overline{\tau}_k$ denotes the complex conjugate of τ_k. If the pencil $\lambda E - A$ is c-stable, then X_k converges towards the solution of the projected Lyapunov equation (2.55). The rate of convergence depends strongly on the choice of the shift parameters. The optimal shift parameters providing the superlinear convergence satisfy the generalized ADI minimax problem

$$
\{\tau_1, \ldots, \tau_q\} = \underset{\{\tau_1, \ldots, \tau_q\} \in \mathbb{C}_-}{\arg\min} \; \underset{t \in \mathrm{Sp}_f(E,A)}{\max} \frac{|(1 - \overline{\tau}_1 t) \cdot \ldots \cdot (1 - \overline{\tau}_q t)|}{|(1 + \tau_1 t) \cdot \ldots \cdot (1 + \tau_q t)|},
$$

where $\mathrm{Sp}_f(E, A)$ denotes the finite spectrum of the pencil $\lambda E - A$. If the matrices E and A satisfy the symmetry condition (2.24), then $\lambda E - A$ has real non-positive eigenvalues. In this case, the optimal real shift parameters can be determined by the selection procedure proposed in [78] once the spectral bounds

$$a = \min\{\lambda_k : \lambda_k \in \mathrm{Sp}_-(E,A)\}, \qquad b = \max\{\lambda_k : \lambda_k \in \mathrm{Sp}_-(E,A)\}$$

are available. Here $\mathrm{Sp}_-(E, A)$ denotes the set of finite eigenvalues of $\lambda E - A$ with negative real part. In general case, the suboptimal ADI parameters can be obtained from a set of largest and smallest in modulus approximate finite eigenvalues of $\lambda E - A$ computed by an Arnoldi procedure [50, 70]. Other parameter selection techniques developed for standard Lyapunov equations [15, 63, 77] can also be used for the projected Lyapunov equation (2.55).

A low-rank approximation to the solution of the projected Lyapunov equation (2.55) can be computed in factored form $X \approx Z_k Z_k^T$ using a low-rank version of the ADI method (LR-ADI) as presented in Algorithm 2.6.

In order to guarantee for the factors Z_k to be real in case of complex shift parameters, we take these parameters in complex conjugate pairs $\{\tau_k, \tau_{k+1} = \overline{\tau}_k\}$.

Algorithm 2.6 The LR-ADI method for the projected Lyapunov equation

---

Given $E, A \in \mathbb{R}^{n,n}$, $B \in \mathbb{R}^{n,m}$, projector P_l and shift parameters $\tau_1, \ldots, \tau_q \in \mathbb{C}_-$, compute a low-rank approximation $X \approx Z_k Z_k^T$ to the solution of the projected Lyapunov equation (2.55).

1. $Z^{(1)} = \sqrt{-2\mathrm{Re}(\tau_1)}\,(E + \tau_1 A)^{-1} P_l B, \qquad Z_1 = Z^{(1)};$
2. FOR $\quad k = 2, 3, \ldots$

$$Z^{(k)} = \sqrt{\frac{\mathrm{Re}(\tau_k)}{\mathrm{Re}(\tau_{k-1})}}\left(I - (\overline{\tau}_{k-1} + \tau_k)(E + \tau_k A)^{-1} A\right) Z^{(k-1)},$$

$$Z_k = [Z_{k-1}, \; Z^{(k)}];$$

END FOR

Then a novel approach for efficient handling of complex shift parameters in the LR-ADI method developed in [16] can also be extended to the projected Lyapunov equation (2.55). At each ADI iteration we have $Z_k = [Z^{(1)}, \ldots, Z^{(k)}] \in \mathbb{R}^{n,mk}$. To keep the low-rank structure in Z_k for large mk, we can compress the columns of Z_k using the rank-revealing QR factorization [18] as described in [9].

Finally, note that the matrices $(E + \tau_k A)^{-1}$ in Algorithm 2.6 do not have to be computed explicitly. Instead, we solve linear systems of the form $(E + \tau_k A)x = P_l b$ either by computing (sparse) LU factorizations and forward/backward substitutions or by using iterative Krylov subspace methods [62].

2.5.2 Newton's Method for Projected Riccati Equations

We consider now the numerical solution of the projected Riccati equation

$$EXF^T + FXE^T + EXB_c^T B_c XE^T + P_l B_o B_o^T P_l^T = 0, \quad X = P_r X P_r^T \tag{2.57}$$

with F, B_c and B_o as in (2.22). Such an equation has to be solved in Algorithm 2.1. The minimal solution $X_{\min}$ of (2.57) is at least *semi-stabilizing* in the sense that all the finite eigenvalues of $\lambda E - F - EX_{\min}B_c^T B_c$ are in the closed left half-plane.

Consider the spaces

$$\mathbb{S}_{P_r} = \{X \in \mathbb{R}^{n,n} : X = X^T, \; X = P_r X P_r^T\},$$
$$\mathbb{S}_{P_l} = \{X \in \mathbb{R}^{n,n} : X = X^T, \; X = P_l X P_l^T\}.$$

Since $X = P_r X P_r^T$, $EP_r = P_l E$ and $FP_r = P_l F$, the Riccati operator given by

$$\mathscr{R}(X) = EXF^T + FXE^T + EXB_c^T B_c XE^T + P_l B_o B_o^T P_l^T$$

maps $\mathbb{S}_{P_r}$ into $\mathbb{S}_{P_l}$. Then the Frechét derivative of $\mathscr{R}$ at $X \in \mathbb{S}_{P_r}$ is a linear operator $\mathscr{R}'_X : \mathbb{S}_{P_r} \to \mathbb{S}_{P_l}$ defined as

$$\mathscr{R}'_X(N) = \lim_{\delta \to 0} \frac{1}{\delta} \big(\mathscr{R}(X + \delta N) - \mathscr{R}(X) \big)$$

for $N \in \mathbb{S}_{P_r}$. Taking into account that $N = P_r N = N P_r^T$, we have

$$\mathscr{R}'_X(N) = (F + EXB_c^T B_c) N E^T + EN(F + EXB_c^T B_c)^T.$$

Then *Newton's method* for the projected Riccati equation (2.57) can be written as

$$\begin{aligned} N_j &= -(\mathscr{R}'_{X_j})^{-1}(\mathscr{R}(X_j)), \\ X_{j+1} &= X_j + N_j. \end{aligned}$$

The standard formulation of this method is given in Algorithm 2.7.

As in the standard case [41], we can combine the second and third steps in Algorithm 2.7 and compute the new iterate X_{j+1} directly from the projected Lyapunov equation as presented in Algorithm 2.8.

Algorithm 2.7 Newton's method for the projected Riccati equation

---

Given $E, F \in \mathbb{R}^{n,n}, B_c \in \mathbb{R}^{m,n}, B_o \in \mathbb{R}^{n,m}$, projectors P_r, P_l and a stabilizing initial guess X_0, compute an approximate solution of the projected Riccati equation (2.57).

FOR $j = 0, 1, 2, \ldots$

1. *Compute $F_j = F + EX_j B_c^T B_c$.*
2. *Solve the projected Lyapunov equation*

$$F_j N_j E^T + EN_j F_j^T = -P_l \mathscr{R}(X_j) P_l^T, \quad N_j = P_r N_j P_r^T.$$

3. *Compute $X_{j+1} = X_j + N_j$.*

END FOR

Algorithm 2.8 The Newton-Kleinman method for the projected Riccati equation

---

Given $E, F \in \mathbb{R}^{n,n}, B_c \in \mathbb{R}^{m,n}, B_o \in \mathbb{R}^{n,m}$, projectors P_r, P_l and a stabilizing initial guess X_0, compute an approximate solution of the projected Riccati equation (2.57).

FOR $j = 1, 2, \ldots$

1. *Compute $K_j = EX_{j-1} B_c^T$ and $F_j = F + K_j B_c$.*
2. *Solve the projected Lyapunov equation*

$$EX_j F_j^T + F_j X_j E^T = -P_l (B_o B_o^T - K_j K_j^T) P_l^T, \qquad X_j = P_r X_j P_r^T.$$

END FOR

Although Algorithms 2.7 and 2.8 are equivalent, they behave different in finite precision arithmetic and there are significant differences in their implementation especially for large-scale problems.

Similarly to the standard state space case [8, 74], one can show that if $\lambda E - F$ is c-stable, then for $X_0 = 0$, all $\lambda E - F_j$ are also c-stable and $\lim_{j \to \infty} X_j = X_{\min}$, see [10]. The convergence is quadratic if the pencil $\lambda E - F - EX_{\min}B_c^T B_c$ is c-stable. Some difficulties may occur if the pencil $\lambda E - F$ has eigenvalues on the imaginary axis. For circuit equations, these eigenvalues are uncontrollable and unobservable [54]. In that case, similarly to [12], one could choose a special stabilizing initial guess X_0 that ensures the convergence of the Newton-Kleinman iteration. However, the computation of such a guess for large-scale problems remains an open problem.

A low-rank approximation to the minimal solution of the projected Riccati equation (2.57) can be computed in factored form $X_{\min} \approx \tilde{R}\tilde{R}^T$ with $\tilde{R} \in \mathbb{R}^{n,k}$, $k \ll n$ using the same approach as in [11]. Starting with $K_1 = EX_0B_c^T$ and $F_1 = F + K_1B_c$, we solve in each Newton-Kleinman iteration two projected Lyapunov equations

$$EX_{1,j}F_j^T + F_jX_{1,j}E^T = -P_lB_oB_o^TP_l^T, \qquad X_{1,j} = P_rX_{1,j}P_r^T, \qquad (2.58)$$

$$EX_{2,j}F_j^T + F_jX_{2,j}E^T = -P_lK_jK_j^TP_l^T, \qquad X_{2,j} = P_rX_{2,j}P_r^T, \qquad (2.59)$$

for the low-rank approximations $X_{1,j} \approx R_{1,j}R_{1,j}^T$ and $X_{2,j} \approx R_{2,j}R_{2,j}^T$, respectively, and then compute $K_{j+1} = E(R_{1,j}R_{1,j}^T - R_{2,j}R_{2,j}^T)B_c^T$ and $F_{j+1} = F + K_{j+1}B_c$. If the convergence is observed after $j_{\max}$ iterations, then an approximate solution $X_{\min} \approx \tilde{R}\tilde{R}^T$ of the projected Riccati equation (2.57) can be computed in factored form by solving the projected Lyapunov equation

$$EXF^T + FXE^T = -P_lQQ^TP_l^T, \qquad X = P_rXP_r^T \qquad (2.60)$$

with $Q = [B_o, \; E(X_{1,j_{\max}} - X_{2,j_{\max}})B_c^T]$ provided $\lambda E - F$ is c-stable. For computing low-rank factors of the solutions of the projected Lyapunov equations (2.58)–(2.60), we can use the generalized LR-ADI method presented in Sect. 2.5.1. Note that in this method we need to compute the products $(E + \tau F_j)^{-1}w$, where $w \in \mathbb{R}^n$ and $E + \tau F_j = E + \tau(A - BB^T) - \tau\hat{K}_jB_c$ with the low-rank matrices $B_c \in \mathbb{R}^{m,n}$ and $\hat{K}_j = \sqrt{2}P_lBM_0^TJ_c^{-T} - K_j \in \mathbb{R}^{n,m}$. One can use the Sherman-Morrison-Woodbury formula [28, Sect. 2.1.3] to compute these products as

$$(E + \tau F_j)^{-1}w = w_1 + M_{\hat{K}_j}\Big((I_m - B_cM_{\hat{K}_j})^{-1}B_cw_1\Big),$$

where $w_1 = (E + \tau(A - BB^T))^{-1}w$ and $M_{\hat{K}_j} = \tau(E + \tau(A - BB^T))^{-1}\hat{K}_j$ can be determined by solving linear systems with the sparse matrix $E + \tau(A - BB^T)$ either by computing sparse LU factorization or by using Krylov subspace methods [62].

2.6   MATLAB Toolbox PABTEC

In this section, we briefly describe the MATLAB toolbox PABTEC which provides some functions for analysis, model reduction, simulation and visualization of circuit equations. PABTEC stays for PAssivity-preserving Balanced Truncation for Electrical Circuits.

Figure 2.3 shows the main strategy of PABTEC. First, the user has to specify the electrical circuit under consideration. The input data for the main routine pabtec.m are incidence matrices describing the topological structure of the circuit, element matrices for linear circuit components, element relations for nonlinear circuit components and some parameters that can be initialized and verified in the routine inipar.m.

Once the circuit is specified, it can be analyzed with the PABTEC routine sysana.m which delivers the information on the topology structure of the circuit, well-posedness and index. If the circuit contains nonlinear elements, it will be decoupled into linear and nonlinear parts as described in Sect. 2.4. Model reduction of the linear (sub)circuit is implemented in the routine pabtecl.m. Linear circuits that contain neither inductors nor capacitors and circuits without resistors cannot be reduced with PABTEC. For model reduction of large resistive network, one

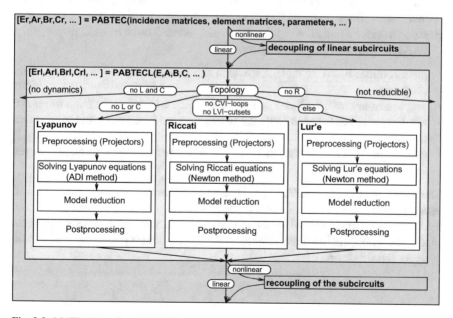

Fig. 2.3 MATLAB toolbox PABTEC

can use a graph-based algorithm presented in [60], which is not yet included in PABTEC. For other network structures, an appropriate model reduction method will be chosen automatically. RC and RL circuits are reduced by the Lyapunov-based balanced truncation algorithms presented in Sect. 2.3.2, while for model reduction of general RLC circuit, Algorithm 2.1 is applied. If the circuit does not contain CVI-loops and LIV-cutsets, then the Gramian in this algorithm is determined by solving the projected Riccati equation (2.57). Otherwise, we have to solve the projected Lur'e equation (2.15). Independent of the topological structure of the circuit, the model reduction algorithms include the following steps: preprocessing, solving matrix equations, model reduction and postprocessing. In the preprocessing step, the basis matrices required in the projector P_r or P are computed using graph-theoretical algorithms. If necessary, also the auxiliary matrices H_k and the matrix M_0 are computed. The projected Lyapunov equations are solved by the LR-ADI method described in Sect. 2.5.1, while the projected Riccati equation is solved by the Newton or Newton-Kleinman method presented in Sect. 2.5.2. The numerical methods for large-scale projected Lur'e equations proposed in [53] will be included in a future release of PABTEC. Note that the matrix equation solvers in PABTEC can be seen as extension of the corresponding functions in the MATLAB toolbox LyaPACK[1][51] and its successor MESS.[2] Postprocessing involves the computation of error bounds and reduced-order initial vectors required for simulation.

The last step in PABTEC is combination of the reduced-order linear model with unchanged nonlinear part. PABTEC includes also the routines `simorih.m` and `simredh.m` for simulation of the original and reduced-order models, respectively.

PABTEC provides the following routines:

- main routines listed in Table 2.1, which can be called by the user in the main program;
- supplementary routines listed in Table 2.2, which are used in the main routines;
- auxiliary routines listed in Table 2.3 which are used in the main and supplementary routines;
- user supplied routines listed in Table 2.4, which provide the information on the circuit.

The MATLAB toolbox PABTEC can be used by a line command or via a graphical user interface (GUI). PABTEC-GUI contains four tab panels shown in Figs. 2.4 and 2.5 that enable the user to upload the circuit, set up all required parameters, perform model reduction, simulate the original and reduced-order systems and save these systems and simulation data.

---

[1] http://www.netlib.org/lyapack/.

[2] https://www.mpi-magdeburg.mpg.de/projects/mess

Table 2.1 Main subroutines in PABTEC

Subroutines	Description
System analysis and parameter initialization	
inipar	Initialization of the parameters for other main subroutines
sysana	Analyzing the topological structure, well-posedness, index of the MNA system
Model order reduction	
pabtec	Model reduction of nonlinear circuit equations via decoupling linear and nonlinear parts and reduction the linear subsystem, see Sect. 2.4
pabtecl	Model reduction of linear circuit equations using the passivity-preserving balanced truncation methods presented in Sect. 2.3
pabtecgui	Graphical user interface for PABTEC
Simulation	
simorih	Simulation of the original nonlinear system using the BDF method with h-scaling of the algebraic equations
simredh	Simulation of the reduced-order nonlinear system using the BDF method with h-scaling of the algebraic equations
Visualization	
plotgraph	Plot a circuit graph

Table 2.2 Supplementary subroutines in PABTEC

Subroutines	Description
Model reduction	
bt_rcl	Balanced truncation for RLC circuits, see Algorithm 2.1
bt_rci	Balanced truncation for RCI circuits, see Algorithm 2.2
bt_rcv	Balanced truncation for RCV circuits, see Algorithm 2.3
bt_rciv	Balanced truncation for RCIV circuits, see Algorithm 2.4
bt_rcvi	Balanced truncation for RCIV circuits, see Algorithm 2.5
bt_rli	Balanced truncation for RLI circuits
bt_rlv	Balanced truncation for RLV circuits
bt_rliv	Balanced truncation for RLIV circuits
bt_rlvi	Balanced truncation for RLIV circuits
Lyapunov equations	
glradi	LR-ADI method for the projected Lyapunov equation, see Algorithm 2.6
glradis	LR-ADI method for the projected symmetric Lyapunov equation
gpar	Computing the suboptimal ADI parameters
gparsym	Computing the suboptimal ADI parameters for symmetric problem
gpar_wach	Computing the optimal ADI shift parameters for symmetric problem
Riccati equations	
glrnw	Low-rank Newton method for the projected Riccati equation, see Algorithm 2.7
glrnwkl	Low-rank Newton-Kleinman method for the projected Riccati equation, see Algorithm 2.8

Table 2.3 Auxiliary subroutines in PABTEC

Subroutines	Description
Graph-theoretical algorithms	
fastlists	Forming a node-branch-list from an incidence matrix
forest	Computing a forest in the graph
inc_bas	Computing the basis matrices for the left null and range spaces of an incidence matrix
inc_rank	Computing the rank of an incidence matrix
loopmatr	Computing a reduced loop matrix from a reduced incidence matrix
Numerical linear algebra	
garn	Arnoldi method for computing the largest and smallest finite eigenvalues of a pencil
garnsym	Arnoldi method for computing the largest and smallest finite eigenvalues of a symmetric pencil
nresl	Computing the residual norms for the projected Lyapunov equation using updated QR factorizations
nresr	Computing the residual norm for the projected Riccati equation using QR factorization
prodinvsym	Computing the matrix-vector product $E^- v$, where E^- is a reflexive inverse of symmetric E w.r.t. to a projector P
prodp	Computing the matrix-vector product $E^- v$, where E^- is a reflexive inverse of E w.r.t. to the projectors P_r and P_l
prodpsym	Computing the projector-vector product Pv
prodpl	Computing the projector-vector product $P_l v$ or $P_l^T v$
prodpr	Computing the projector-vector product $P_r v$ or $P_r^T v$
Miscellaneous	
hmatr	Computing the matrices H_k required for the projectors P_r and P_l
incmat	Determination of the incidence matrices
ininet	Initialization of the network topology
matr2ascii	Export the matrix in ASCII-format
m0m	Computing the matrix M_0 as in (2.40)
mnadae	Construction of E, A, B from incidence and element matrices

Table 2.4 User supplied subroutines for PABTEC

Subroutines	Description
netlist	Incidence and element matrices for linear circuit components
PCN	Matrix valued function for nonlinear capacitors
PLN	Matrix valued function for nonlinear inductors
gN	Nonlinear current-voltage relation for nonlinear resistors
uV	Voltages of voltage sources
iI	Currents of current sources

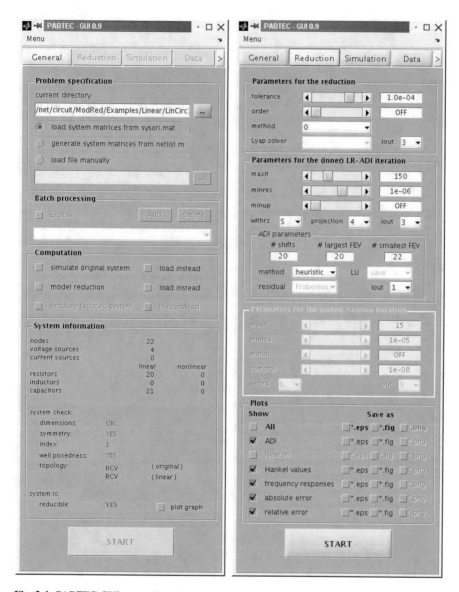

Fig. 2.4 PABTEC-GUI: general panel

2.7   Numerical Examples

In this section, we present some results of numerical experiments to demonstrate the properties of the presented model reduction methods for linear and nonlinear circuits. The computations were done on IBM RS 6000 44P Model 270 with machine precision $\varepsilon = 2.22 \times 10^{-16}$ using MATLAB 7.0.4.

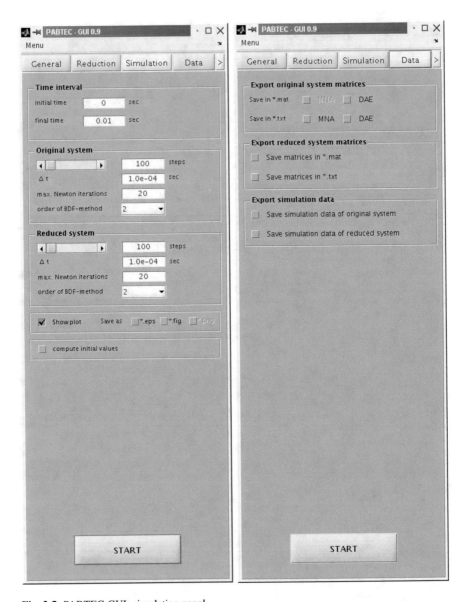

Fig. 2.5 PABTEC-GUI: simulation panel

Example 2.7.1 The first example is a transmission line model from [7] consisting of a scalable number of RLC ladders. We have a single-input-single-output reciprocal passive DAE system (2.9), (2.10) of order $n = 60,000$. The minimal solution of the projected Riccati equation (2.21) was approximated by a low-rank matrix $X_{\min} \approx \widetilde{R}\widetilde{R}^T$ with $\widetilde{R} \in \mathbb{R}^{n,58}$ using Newton's method as presented in Algorithm 2.7.

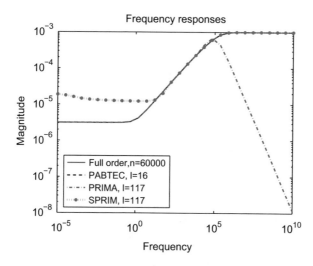

Fig. 2.6 Transmission line: the frequency responses of the original system and the reduced-order models computed by the PABTEC, PRIMA and SPRIM methods

The original system was approximated by a reduced model of order $n_r = 16$ using Algorithm 2.1 with the error bound $\gamma = 2.72 \cdot 10^{-5}$, where

$$\gamma = \frac{\|I + \hat{\mathbf{G}}\|_{\mathcal{H}_\infty}^2 (\gamma_{r+1} + \ldots + \gamma_q)}{1 - \|I + \hat{\mathbf{G}}\|_{\mathcal{H}_\infty} (\gamma_{r+1} + \ldots + \gamma_q)}, \qquad r = 15.$$

For comparison, we have also computed the reduced models of order $n_r = 117$ using the PRIMA and SPRIM algorithms [26, 48]. This order was chosen as a smallest integer such that the absolute error $|\hat{\mathbf{G}}(j\omega) - \mathbf{G}(j\omega)|$ for the SPRIM model is below γ on the frequency interval $[10^{-5}, 10^{10}]$. In Fig. 2.6, we display the magnitude of the frequency responses $\mathbf{G}(j\omega)$ and $\hat{\mathbf{G}}(j\omega)$ of the original and reduced-order models. Figure 2.7 shows the absolute errors $|\hat{\mathbf{G}}(j\omega) - \mathbf{G}(j\omega)|$ and also the error bound γ. One can see that PABTEC provides a much smaller system with keeping the better global approximation properties. It should also be noted that the result for SPRIM is presented here for the best choice of the expansion point that was found after several runs of this algorithm. Taking this into account, the reduction time for the PABTEC method becomes comparable to the actual reduction time for SPRIM. ◁

Example 2.7.2 Next we consider a nonlinear circuit shown in Fig. 2.8. It contains 1501 linear capacitors, 1500 linear resistors, 1 voltage source and 1 diode. Such a circuit is described by the DAE system (2.2), (2.3) of the state space dimension $n = 1503$. We simulate this system on the time interval $\mathbb{I} = [0\,\text{s}, 0.07\,\text{s}]$ with a fixed stepsize 10^{-5} s using the BDF method of order 2. The voltage source is given by $v_\mathcal{V}(t) = 10 \sin(100\pi t)^4$ V, see Fig. 2.9. The linear resistors have the same resistance

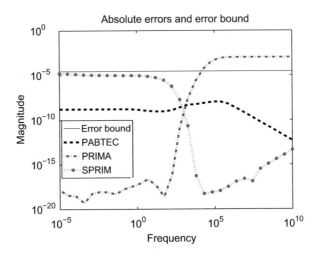

Fig. 2.7 Transmission line: the absolute errors and the error bound (2.20)

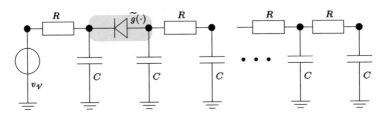

Fig. 2.8 Nonlinear RC circuit

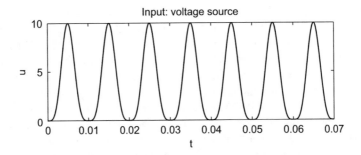

Fig. 2.9 Voltage source for the RC circuit

$R = 2 \, \text{k}\Omega$, the linear capacitors have the same capacitance $C = 0.02 \, \mu\text{F}$ and the diode has a characteristic curve $g(v_{\tilde{\mathcal{R}}}) = 10^{-14}(\exp(40\frac{1}{V}v_{\tilde{\mathcal{R}}}) - 1) \, \text{A}$.

The diode was replaced by an equivalent linear circuit as described in Sect. 2.4. The resulting linear system of order $n_\ell = 1504$ was approximated by a reduced model of order $n_r = r + r_0$, where $r_0 = \text{rank}(I - M_0)$ and r satisfies the condition $(\gamma_{r+1} + \ldots + \gamma_q) < tol$ with a prescribed tolerance tol. For comparison, we

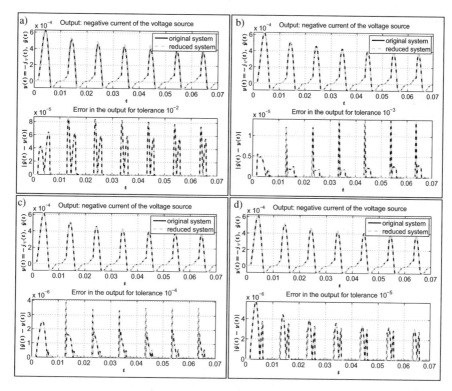

Fig. 2.10 Outputs of the original and the reduced-order nonlinear systems and the errors in the output for the different tolerances (**a**) 10^{-2}, (**b**) 10^{-3}, (**c**) 10^{-4}, (**d**) 10^{-5}

compute the reduced-order linear models for the different tolerances $tol = 10^{-2}$, $10^{-3}, 10^{-4}, 10^{-5}$. The numerical results are given in Fig. 2.10. In the upper plot of each subfigure, we present the computed outputs $y(t) = -j_V(t)$ and $\hat{y}(t)$ of the original and reduced-order nonlinear systems, respectively, whereas the lower plot shows the error $|\hat{y}(t) - y(t)|$.

Table 2.5 demonstrates the efficiency of the proposed model reduction method. One can see that for the decreasing tolerance, the dimension of the reduced-order system increases while the error in the output decreases. The speedup is defined as the simulation time for the original system divided by the simulation time for the reduced-order model. For example, a speedup of 219 in simulation of the reduced-order nonlinear model of dimension $\hat{n} = 13$ with the error $\|\hat{y} - y\|_{L_2(\mathbb{I})} = 6.2 \cdot 10^{-7}$ was achieved compared to the simulation of the original system.   ◁

Example 2.7.3 We consider now the nonlinear circuit shown in Fig. 2.11. It contains 1000 repetitions of subcircuits consisting of one inductor, two capacitors and two resistors. Furthermore, at the beginning and at the end of the chain, we have a voltage source with $v_V(t) = \sin(100\pi t)^{10}$ V as in Fig. 2.12 and an additional

Table 2.5 Statistics for the RC circuit

Dimension of the original nonlinear system, n	1503	1503	1503	1503
Simulation time for the original system, t_{sim}	24,012 s	24,012 s	24,012 s	24,012 s
Tolerance for model reduction of the linear subsystem, tol	1e-02	1e-03	1e-04	1e-05
Time for model reduction, t_{mor}	15 s	24 s	42 s	61 s
Dimension of the reduced nonlinear system, $\hat{n}$	10	13	16	19
Simulation time for the reduced system, $\hat{t}_{\text{sim}}$	82 s	110 s	122 s	155 s
Error in the output, $\|\hat{y} - y\|_{L_2(\mathbb{I})}$	7.0e-06	6.2e-07	2.0e-07	4.2e-07
Speedup, $t_{\text{sim}}/\hat{t}_{\text{sim}}$	294.0	219.0	197.4	155.0

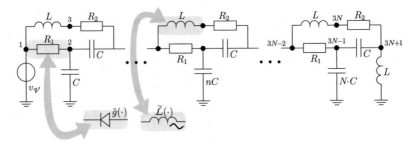

Fig. 2.11 Nonlinear RLC circuit

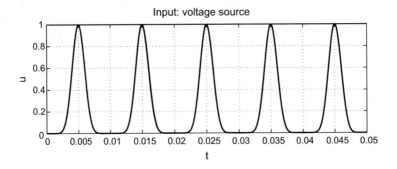

Fig. 2.12 Voltage source for RLC circuit

linear inductor, respectively. In the 1st, 101st, 201st, etc., subcircuits, a linear resistor is replaced by a diode, and in the 100th, 200th, 300th, etc., subcircuits, a linear inductor is replaced by a nonlinear inductor. The resulting nonlinear circuit contains one voltage source, 1990 linear resistors with $R_1 = 20\,\Omega$ and $R_2 = 1\,\Omega$, 991 linear inductors with $L = 0.01\,\text{H}$, 2000 linear capacitors with $C = 1\,\mu\text{F}$, ten diodes with $\widetilde{g}(v_{\widetilde{\mathcal{R}}}) = 10^{-14}(\exp(40\frac{1}{V}v_{\widetilde{\mathcal{R}}}) - 1)\,\text{A}$, and ten nonlinear inductors with

$$\widetilde{L}(j_{\widetilde{L}}) = L_{\min} + (L_{\max} - L_{\min})\exp(-j_{\widetilde{L}}^2 L_{\text{scl}}),$$

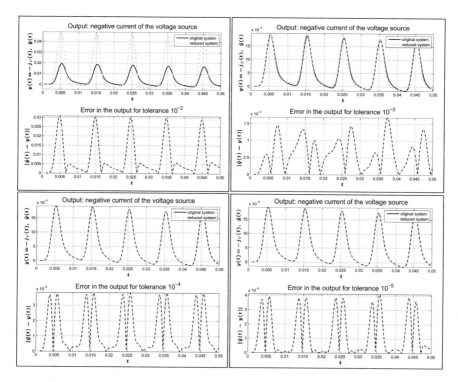

Fig. 2.13 The outputs of the original and the reduced-order nonlinear systems and the errors in the output for the different tolerances (**a**) 10^{-2}, (**b**) 10^{-3}, (**c**) 10^{-4}, (**d**) 10^{-5}

where $L_{\min} = 0.001$ H, $L_{\max} = 0.002$ H and $L_{\mathrm{scl}} = 10^4 \frac{1}{\mathrm{A}}$. The state space dimension of the resulting DAE system is $n = 4003$.

The numerical simulation is done on the time interval $\mathbb{I} = [0\,\mathrm{s}, 0.05\,\mathrm{s}]$ using the BDF method of order 2 with a fixed stepsize of length $5 \cdot 10^{-5}$ s. In Fig. 2.13, we again present the outputs $y(t) = -j_V(t)$ and $\hat{y}(t)$ of the original and reduced-order nonlinear systems, respectively, as well as the error $|\hat{y}(t) - y(t)|$ for the different tolerances $tol = 10^{-2}, 10^{-3}, 10^{-4}, 10^{-5}$ for model reduction of the decoupled linear subcircuit. Table 2.6 demonstrates the efficiency of the model reduction method. As in the example above, also here one can see that if the tolerance decreases, the dimension of the reduced-order system increases while the error in the output becomes smaller. In particular, for the approximate nonlinear model of dimension $\hat{n} = 189$ with the error $\|\hat{y} - y\|_{\mathbb{L}_2(\mathbb{I})} = 4.10 \cdot 10^{-5}$, the simulation time is only 57 s instead of 1 h and 13 min for the original system that implies a speedup of 76.8.

Other results of numerical experiments with PABTEC can be found in Chaps. 1, 4 and 5 in this book.

◁

Table 2.6 Statistics for the RLC circuit

Dimension of the original nonlinear system, n	4003	4003	4003	4003
Simulation time for the original system t_{sim}	4390 s	4390 s	4390 s	4390 s
Tolerance for model reduction of the linear subsystem, tol	1e-02	1e-03	1e-04	1e-05
Time for the model reduction, t_{mor}	2574 s	2598 s	2655 s	2668 s
Dimension of the reduced nonlinear system, $\hat{n}$	127	152	189	218
Simulation time for the reduced system, $\hat{t}_{sim}$	33 s	42 s	57 s	74 s
Error in the output, $\|\hat{y} - y\|_{L_2(\mathbb{I})}$	2.73e-03	1.67e-04	4.10e-05	4.09e-05
Speedup, $t_{sim}/\hat{t}_{sim}$	132.0	104.1	76.8	59.1

Acknowledgements The work reported in this paper was supported by the German Federal Ministry of Education and Research (BMBF), grant no. 03STPAE3. Responsibility for the contents of this publication rests with the authors.

References

1. Anderson, B., Vongpanitlerd, S.: Network Analysis and Synthesis. Prentice Hall, Englewood Cliffs, NJ (1973)
2. Andrásfai, B.: Graph Theory: Flows, Matrices. Adam Hilger, New York/Bristol (1991)
3. Antoulas, A.: Approximation of Large-Scale Dynamical Systems. SIAM, Philadelphia, PA (2005)
4. Antoulas, A.: A new result on passivity preserving model reduction. Syst. Control Lett. **54**(4), 361–374 (2005)
5. Antoulas, A., Beattie, C., Gugercin, S.: Interpolatory model reduction of large-scale dynamical systems. In: Mohammadpour, J., Grigoriadis, K. (eds.) Efficient Modeling and Control of Large-Scale Systems, pp. 3–58. Springer, New York (2010)
6. Bai, Z.: Krylov subspace techniques for reduced-order modeling of large-scale dynamical systems. Appl. Numer. Math. **43**, 9–44 (2002)
7. Bechtold, T., Verhoeven, A., ter Maten, E., Voss, T.: Model order reduction: an advanced, efficient and automated computational tool for microsystems. In: Cutello, V., Fotia, G., Puccio, L. (eds.) Applied and Industrial Mathematics in Italy II, Selected Contributions from the 8th SIMAI Conference. Advances in Mathematics for Applied Sciences, vol. 75, pp. 113–124. World Scientific, Singapore (2007)
8. Benner, P.: Numerical solution of special algebraic Riccati equations via exact line search method. In: Proceedings of the European Control Conference (ECC97), paper 786. BELWARE Information Technology, Waterloo (1997)
9. Benner, P., Quintana-Ortí, E.: Solving stable generalized Lyapunov equations with the matrix sign function. Numer. Algorithms **20**(1), 75–100 (1999)
10. Benner, P., Stykel, T.: Numerical solution of projected algebraic Riccati equations. SIAM J. Numer. Anal. **52**(2), 581–600 (2014)
11. Benner, P., Quintana-Ortí, E., Quintana-Ortí, G.: Efficient numerical algorithms for balanced stochastic truncation. Int. J. Appl. Math. Comput. Sci. **11**(5), 1123–1150 (2001)
12. Benner, P., Hernández, V., Pastor, A.: The Kleinman iteration for nonstabilizable systems. Math. Control Signals Syst. **16**, 76–93 (2003)

13. Benner, P., Mehrmann, V., Sorensen, D. (eds.): Dimension Reduction of Large-Scale Systems. Lecture Notes in Computational Science and Engineering, vol. 45. Springer, Berlin/Heidelberg (2005)
14. Benner, P., Li, J.R., Penzl, T.: Numerical solution of large Lyapunov equations, Riccati equations, and linear-quadratic control problems. Numer. Linear Algebra Appl. **15**, 755–777 (2008)
15. Benner, P., Mena, H., Saak, J.: On the parameter selection problem in the Newton-ADI iteration for large-scale Riccati equations. Electron. Trans. Numer. Anal. **29**, 136–149 (2008)
16. Benner, P., Kürschner, P., Saak, J.: Efficient handling of complex shift parameters in the low-rank Cholesky factor ADI method. Numer. Algorithms **62**(2), 225–251 (2013)
17. Brenan, K., Campbell, S., Petzold, L.: The Numerical Solution of Initial-Value Problems in Differential-Algebraic Equations. Classics in Applied Mathematics, vol. 14. SIAM, Philadelphia, PA (1996)
18. Chan, T.: Rank revealing QR factorizations. Linear Algebra Appl. **88/89**, 67–82 (1987)
19. Chua, L.: Dynamic nonlinear networks: state-of-the-art. IEEE Trans. Circuits Syst. **27**(11), 1059–1087 (1980)
20. Chua, L., Desoer, C., Kuh, E.: Linear and Nonlinear Circuits. McGraw-Hill, New York (1987)
21. Dai, L.: Singular Control Systems. Lecture Notes in Control and Information Sciences, vol. 118. Springer, Berlin/Heidelberg (1989)
22. Deo, N.: Graph Theory with Application to Engineering and Computer Science. Prentice-Hall, Englewood Cliffs, NJ (1974)
23. Estévez Schwarz, D.: A step-by-step approach to compute a consistent initialization for the MNA. Int. J. Circuit Theory Appl. **30**, 1–16 (2002)
24. Estévez Schwarz, D., Tischendorf, C.: Structural analysis for electric circuits and consequences for MNA. Int. J. Circuit Theory Appl. **28**, 131–162 (2000)
25. Freund, R.: Model reduction methods based on Krylov subspaces. Acta Numer. **12**, 267–319 (2003)
26. Freund, R.: SPRIM: structure-preserving reduced-order interconnect macromodeling. In: Technical Digest of the 2004 IEEE/ACM International Conference on Computer-Aided Design, pp. 80–87. IEEE Computer Society, Los Alamos, CA (2004)
27. Freund, R.: Structure-preserving model order reduction of RCL circuit equations. In: Schilders, W., van der Vorst, H., Rommes, J. (eds.) Model Order Reduction: Theory, Research Aspects and Applications. Mathematics in Industry, vol. 13, pp. 49–73. Springer, Berlin/Heidelberg (2008)
28. Golub, G., Van Loan, C.: Matrix Computations, 3rd edn. The Johns Hopkins University Press, Baltimore/London (1996)
29. Griepentrog, E., März, R.: Differential-Algebraic Equations and Their Numerical Treatment. Teubner-Texte zur Mathematik, vol. 88. B.G. Teubner, Leipzig (1986)
30. Grimme, E.: Krylov projection methods for model reduction. Ph.D. thesis, University of Illinois, Urbana-Champaign (1997)
31. Gugercin, S., Antoulas, A.: A survey of model reduction by balanced truncation and some new results. Int. J. Control **77**(8), 748–766 (2004)
32. Gugercin, S., Antoulas, A., Beattie, C.: $\mathcal{H}_2$ model reduction for large-scale linear dynamical systems. SIAM J. Matrix Anal. Appl. **30**(2), 609–638 (2008)
33. Hairer, E., Wanner, G.: Solving Ordinary Differential Equations II - Stiff and Differential-Algebraic Problems, 2nd edn. Springer, Berlin (1996)
34. Heinkenschloss, M., Reis, T.: Model reduction for a class of nonlinear electrical circuits by reduction of linear subcircuits. Technical Report 702–2010, DFG Research Center MATHEON, Technische Universität Berlin (2010). http://http://www.math.tu-berlin.de/~reis/Publicat/pr_10_702.pdf
35. Hinze, M., Kunkel, M.: Residual based sampling in pod model order reduction of drift-diffusion equations in parametrized electrical networks. Z. Angew. Math. Mech. **92**(2), 91–104 (2012)

36. Hinze, M., Kunkel, M., Steinbrecher, A., Stykel, T.: Model order reduction of coupled circuit-device systems. Int. J. Numer. Model. Electron. Networks Devices Fields **25**, 362–377 (2012)
37. Ho, C.W., Ruehli, A., Brennan, P.: The modified nodal approach to network analysis. IEEE Trans. Circuits Syst. **22**(6), 504–509 (1975)
38. Ionutiu, R., Rommes, J., Antoulas, A.: Passivity-preserving model reduction using dominant spectral-zero interpolation. IEEE Trans. Comput. Aided Des. Integr. Circuits Syst. **27**(12), 2250–2263 (2008)
39. Ipach, H.: Graphentheoretische Anwendungen in der Analyse elektrischer Schaltkreise. Bachelorarbeit, Universität Hamburg (2013)
40. Jungnickel, D.: Graphs, Network and Algorithms. Springer, Berlin/Heidelberg (2005)
41. Kleinman, D.: On an iterative technique for Riccati equation computations. IEEE Trans. Autom. Control **13**, 114–115 (1968)
42. Knockaert, L., De Zutter, D.: Laguerre-SVD reduced-order modeling. IEEE Trans. Microwave Theory Tech. **48**(9), 1469–1475 (2000)
43. Kunkel, P., Mehrmann, V.: Differential-Algebraic Equations. Analysis and Numerical Solution. EMS Publishing House, Zürich (2006)
44. Li, J.R., White, J.: Low rank solution of Lyapunov equations. SIAM J. Matrix Anal. Appl. **24**(1), 260–280 (2002)
45. Liu, W., Sreeram, V., Teo, K.: Model reduction for state-space symmetric systems. Syst. Control Lett. **34**(4), 209–215 (1998)
46. Moore, B.: Principal component analysis in linear systems: controllability, observability, and model reduction. IEEE Trans. Autom. Control **26**(1), 17–32 (1981)
47. Ober, R.: Balanced parametrization of classes of linear systems. SIAM J. Control Optim. **29**(6), 1251–1287 (1991)
48. Odabasioglu, A., Celik, M., Pileggi, L.: PRIMA: passive reduced-order interconnect macro-modeling algorithm. IEEE Trans. Comput. Aided Des. Integr. Circuits Syst. **17**(8), 645–654 (1998)
49. Ovalekar, V., Narayanan, H.: Fast loop matrix generation for hybrid analysis and a comparison of the sparsity of the loop impedance and MNA impedance submatrices. In: Proceedings of the IEEE International Symposium on Circuits and Systems, ISCAS '92, vol. 4, pp. 1780–1783 (1992)
50. Penzl, T.: A cyclic low-rank Smith method for large sparse Lyapunov equations. SIAM J. Sci. Comput. **21**(4), 1401–1418 (1999/2000)
51. Penzl, T.: LYAPACK Users Guide. Preprint SFB393/00-33, Fakultät für Mathematik, Technische Universität Chemnitz, Chemnitz (2000). Available from http://www.tu-chemnitz.de/sfb393/sfb00pr.html
52. Phillips, J., Daniel, L., Silveira, L.: Guaranteed passive balancing transformations for model order reduction. IEEE Trans. Comput. Aided Des. Integr. Circuits Syst. **22**(8), 1027–1041 (2003)
53. Poloni, F., Reis, T.: A deflation approach for large-scale Lur'e equations. SIAM. J. Matrix Anal. Appl. **33**(4), 1339–1368 (2012)
54. Reis, T., Stykel, T.: PABTEC: Passivity-preserving balanced truncation for electrical circuits. IEEE Trans. Compu. Aided Des. Integr. Circuits Syst. **29**(9), 1354–1367 (2010)
55. Reis, T., Stykel, T.: Positive real and bounded real balancing for model reduction of descriptor systems. Int. J. Control **83**(1), 74–88 (2010)
56. Reis, T., Stykel, T.: Lyapunov balancing for passivity-preserving model reduction of RC circuits. SIAM J. Appl. Dyn. Syst. **10**(1), 1–34 (2011)
57. Rewieński, M.: A trajectory piecewise-linear approach to model order reduction of nonlinear dynamical systems. Ph.D. thesis, Massachusetts Institute of Technology (2003)
58. Riaza, R., Tischendorf, C.: Qualitative features of matrix pencils and DAEs arising in circuit dynamics. Dyn. Syst. **22**, 107–131 (2007)
59. Riaza, R., Tischendorf, C.: The hyperbolicity problem in electrical circuit theory. Math. Methods Appl. Sci. **33**(17), 2037–2049 (2010)

60. Rommes, J., Schilders, W.: Efficient methods for large resistor networks. IEEE Trans. Comput. Aided Des. Integr. Circuits Syst. **29**(1), 28–39 (2010)
61. Roos, J., Costa, L. (eds.): Scientific Computing in Electrical Engineering SCEE 2008. Mathematics in Industry, vol. 14. Springer, Berlin/Heidelberg (2010)
62. Saad, Y.: Iterative Methods for Sparse Linear Systems. PWS Publishing Company, Boston, MA (1996)
63. Sabino, J.: Solution of large-scale Lyapunov equations via the block modified Smith method. Ph.D. thesis, Rice University, Houston (2006)
64. Schilders, W., van der Vorst, H., J., R. (eds.): Model Order Reduction: Theory, Research Aspects and Applications. Mathematics in Industry, vol. 13. Springer, Berlin/Heidelberg (2008)
65. Sirovich, L.: Turbulence and the dynamics of coherent structures. I: coherent structures. II: symmetries and transformations. III: dynamics and scaling. Q. Appl. Math. **45**, 561–590 (1987)
66. Sorensen, D.: Passivity preserving model reduction via interpolation of spectral zeros. Syst. Control Lett. **54**(4), 347–360 (2005)
67. Soto, M.S., Tischendorf, C.: Numerical analysis of DAEs from coupled circuit and semiconductor simulation. Appl. Numer. Math. **53**(2–4), 471–88 (2005)
68. Steinbrecher, A., Stykel, T.: Model order reduction of nonlinear circuit equations. Int. J. Circuits Theory Appl. **41**, 1226–1247 (2013)
69. Stykel, T.: Gramian-based model reduction for descriptor systems. Math. Control Signals Syst. **16**, 297–319 (2004)
70. Stykel, T.: Low-rank iterative methods for projected generalized Lyapunov equations. Electron. Trans. Numer. Anal. **30**, 187–202 (2008)
71. Stykel, T.: Balancing-related model reduction of circuit equations using topological structure. In: Benner, P., Hinze, M., ter Maten, E. (eds.) Model Reduction for Circuit Simulation. Lecture Notes in Electrical Engineering, vol. 74, pp. 53–80. Springer, Berlin/Heidelberg (2011)
72. Stykel, T., Reis, T.: The PABTEC algorithm for passivity-preserving model reduction of circuit equations. In: Proceedings of the 19th International Symposium on Mathematical Theory of Networks and Systems, MTNS 2010, Budapest, 5–9 July 2010, paper 363. ELTE, Budapest (2010)
73. Tischendorf, C.: Coupled systems of differential algebraic and partial differential equations in circuit and device simulation. Habilitation thesis, Humboldt-Universität Berlin (2004)
74. Varga, A.: On computing high accuracy solutions of a class of Riccati equations. Control Theory Adv. Technol. **10**, 2005–2016 (1995)
75. Vlach, J., Singhal, K.: Computer Methods for Circuit Analysis and Design. Van Nostrand Reinhold, New York (1994)
76. Wachspress, E.: Iterative solution of the Lyapunov matrix equation. Appl. Math. Lett. **1**, 87–90 (1988)
77. Wachspress, E.: The ADI minimax problem for complex spectra. In: Kincaid, D., Hayes, L. (eds.) Iterative Methods for Large Linear Systems, pp. 251–271. Academic, San Diego (1990)
78. Wachspress, E.: The ADI Model Problem. Monograph, Windsor, CA (1995)
79. Weinbreg, L., Ruehili, A.: Combined modified loop analysis, modified nodal analysis for large-scale circuits. IBM Research Report RC 10407, IBM Research Devision (1984)

Chapter 3
Reduced Representation of Power Grid Models

Peter Benner and André Schneider

Abstract We discuss the reduction of large-scale circuit equations with many terminals. Usual model order reduction (MOR) methods assume a small number of inputs and outputs. This is no longer the case, e.g., for the power supply network for the functional circuit elements on a chip. Here, the order of inputs/outputs, or *terminals*, is often of the same order as the number of equations. In order to apply classical MOR techniques to these *power grids*, it is therefore mandatory to first perform a *terminal reduction*. In this survey, we discuss several techniques suggested for this task, and develop an efficient numerical implementation of the extended SVD MOR approach for large-scale problems. For the latter, we suggest to use a truncated SVD computed either by the implicitly restarted Arnoldi method or the Jacobi-Davidson algorithm. We analyze this approach regarding stability, passivity, and reciprocity preservation, derive error bounds, and discuss issues arising in the numerical implementation of this method.

3.1 Introduction

As already discussed and motivated in the previous chapters, it is indisputable that model order reduction (MOR) in circuit simulation is absolutely necessary. This chapter treats the reduction of linear subcircuits. MOR of these (parasitic) subsystems is part of the research focus since decades. A lot of approaches

P. Benner (✉)
Max Planck Institute for Dynamics of Complex Technical Systems, Sandtorstr. 1, 39106 Magdeburg, Germany

Technische Universität Chemnitz, Fakultät für Mathematik, Reichenhainer Str. 39/41, 09126 Chemnitz, Germany
e-mail: benner@mpi-magdeburg.mpg.de, benner@mathematik.tu-chemnitz.de

A. Schneider
Max Planck Institute for Dynamics of Complex Technical Systems, Sandtorstr. 1, 39106 Magdeburg, Germany
e-mail: andre.schneider@mpi-magdeburg.mpg.de

© Springer International Publishing AG 2017
P. Benner (ed.), *System Reduction for Nanoscale IC Design*,
Mathematics in Industry 20, DOI 10.1007/978-3-319-07236-4_3

are available (see, e.g., Chap. 2 or [11, 47]), but recently a structural change handicaps the explored algorithms, even makes them inapplicable in some cases. The established approaches assume a relatively small number of input and output connections of these parasitic systems, which is no longer true for some situations nowadays. In these cases, a simulation of the full unreduced model might be much faster than the model reduction step itself such that MOR is not reasonable anymore.

It becomes increasingly important also to model the power supply of the electronic devices such that a significant class of applications in circuit simulation, which violates the assumption above, are power grid networks. In modern multi-layer integrated circuits (ICs) these power grids are own layers, which are responsible for the power supply of functional circuit elements, e.g., transistors. As a consequence, there is a need for high connectivity, which leads to mathematical models with a lot of inputs and outputs, i.e., one I/O-terminal (also called pin) for each supplied element. The development of new MOR methods being applicable to such challenging LTI systems is subject matter of this chapter.

Our goal is therefore to find a concept to compress the input/output matrices in such a way that we get a terminal reduced system with similar behavior. This system realization enables then the use of classical MOR methods as described, e.g., in the previous chapter. Fundamentally, there are two popular types of MOR methods: either we use modal based methods and Krylov subspace (KROM) methods [2, 10, 11, 15, 19, 39] or Hankel norm approximations and balanced truncation (BT) methods [2, 9–11, 38]. No matter which MOR method we use, with the help of the approach to first reduce the terminals and then to reduce the order of the system, it is possible to apply the original I/O-data to a reduced-order model. The detailed procedure is described in this chapter. We illustrate numerical results obtained for the performance of this approach using examples.

In Sect. 3.2 we lay the foundations for the approach. We introduce basic definitions and explain the mathematical systems appearing in this chapter. We highlight different notations and special emphasis is given to the concept of the moments of the transfer function of a system. Additionally, the numerical examples, which play a role throughout the whole chapter, are introduced. In Sect. 3.3 we attend to different approaches tackling MOR of linear systems with a lot of terminals. In [17], a modified conjugate gradient algorithm is suggested to analyze power grids. Also [16, 27] propose rather theoretical ideas, especially in [16] a method for terminal compression is introduced, called SVDMOR, and is the starting point of our work. In [33], an extended version, the ESVDMOR approach, is published, which we explain in detail. Additionally, we analyze the very similar algorithm TermMerg [32], which compresses the terminals by merging them together in a special manner. We also comment on two approaches based on very different ideas. The first approach, SparseRC [25, 26], is based on considering the network as a graph. This opens the door to graph theoretical ideas, such as partitioning and node elimination, which leads to a MOR of the system. The second approach is based on interpolation of measurements [3, 29]. With the help of the Loewner matrix concept we show how to obtain the minimal realization of a reduced model.

As our focus is on ESVDMOR, Sect. 3.4 deals with the characteristics of this approach. Essential properties of the original system, such as stability, passivity, and reciprocity should not be destroyed during the reduction process. The preservation of these properties is an important task in circuit simulation. We introduce basic definitions and prove that the ESVDMOR approach is stability, passivity, and reciprocity preserving under certain conditions following [7]. Also, the analysis of the approximation error derived in [8] is reviewed. We also point out the numerical bottlenecks and explain how to avoid them. Here, the truncated singular value decomposition (SVD) plays an important role. We will basically introduce two approaches to perform a truncated SVD efficiently, building upon [6]. In Sect. 3.5, we conclude this chapter by assessing the theoretical results and the numerical experiments. We give an outlook to problems of interest for future research in the area of MOR for power grids.

3.2 System Description

3.2.1 Basic Definitions

Modeling in the area of circuit simulation, and also in areas such as mechanical, physical, biological and chemical applications, often leads to linear time-invariant (LTI) continuous-time systems of the form

$$\begin{aligned}
E\dot{x}(t) &= Ax(t) + Bu(t), \ Ex(0) = Ex_0, \\
y(t) &= Cx(t) + Du(t).
\end{aligned} \tag{3.1}$$

With reference to circuit simulation we have the following definitions:

- $A \in \mathbb{R}^{n \times n}$ is the resistance matrix,
- $E \in \mathbb{R}^{n \times n}$ is the conductance matrix,
- $B \in \mathbb{R}^{n \times m}$ is the input matrix,
- $C \in \mathbb{R}^{p \times n}$ is the output matrix, and
- $D \in \mathbb{R}^{p \times m}$ is the feed-through term, which in many practical applications is equal to zero.

Furthermore,

- $x(t) \in \mathbb{R}^n$ contains internal state variables, e.g., currents or voltages,
- $u(t) \in \mathbb{R}^m$ is the vector of input variables, and
- $y(t) \in \mathbb{R}^p$ is the output vector.

To ensure a unique solution we also need the initial value $x_0 \in \mathbb{R}^n$. The number of state variables n is also called the order of the system. The number of inputs m and the number of outputs p are not necessarily equal. Unless otherwise noted, we assume the matrix pencil $(\lambda E - A)$ to be regular, i.e., $\det(\lambda E - A) \neq 0$ for at least one $\lambda \in \mathbb{C}$. The matrix E is allowed to be singular. In this case, the system (3.1) consists of a differential-algebraic equation (DAE) in semi-explicit form combined with an output term. Together, the equations in (3.1) form a descriptor system. In this chapter, we will use the notation explained above, but other notations are also commonly found in the literature, e.g.,

$$
\begin{aligned}
C\dot{x}(t) &= -Gx(t) + Bu(t), \quad Cx(0) = Cx_0, \\
y(t) &= Lx(t) + Eu(t).
\end{aligned}
\tag{3.2}
$$

Applying the Laplace transform to (3.1) leads to

$$
\begin{aligned}
(sE\tilde{x}(s) - Ex(0)) &= A\tilde{x}(s) + B\tilde{u}(s), \\
\tilde{y}(s) &= C\tilde{x}(s) + D\tilde{u}(s).
\end{aligned}
$$

After rearranging the DAE and using the initial condition $Ex(0) = Ex_0$, we get

$$
(sE - A)\tilde{x}(s) = Ex_0 + B\tilde{u}(s), \qquad \text{or}
$$

$$
\tilde{x}(s) = (sE - A)^{-1}Ex_0 + (sE - A)^{-1}B\tilde{u}(s),
$$

which we plug into the output term. Assuming $Ex_0 = 0$, this term leads to a direct linear input-output mapping in frequency domain, $\tilde{y}(s) = \big(C(sE - A)^{-1}B + D\big)\tilde{u}(s)$.

Definition 3.2.1 (The Transfer Function) The rational matrix-valued function

$$
G(s) = C(sE - A)^{-1}B + D,
\tag{3.3}
$$

with $s \in \mathbb{C}$ is called *the transfer function* of the continuous-time descriptor system (3.1). If $s = i\omega$, then $\omega \in \mathbb{R}$ is called the *(radial) frequency*.

Note that, for simplicity, we denote $\tilde{y}(s) \equiv y(s)$ and $\tilde{x}(s) \equiv x(s)$. The distinction between the variables $x(t)$ and $x(s)$ should be clear from the context and the differing arguments (t indicating time, s frequency domain). Using the notation (3.2), the transfer function is sometimes described by

$$
H(s) = L(sC + G)^{-1}B + E.
$$

Modified nodal analysis (MNA) modeling of current driven (impedance form) RLC power grid circuits, i.e. the inputs are currents from current sources injected into the external nodes or terminals of the circuit, leads to systems with the following block structure [20]:

$$\begin{bmatrix} E_1 & 0 \\ 0 & E_2 \end{bmatrix} \dot{x} = \begin{bmatrix} A_1 & A_2 \\ -A_2^T & 0 \end{bmatrix} x + \begin{bmatrix} B_1 \\ 0 \end{bmatrix} u,$$

$$y = \begin{bmatrix} B_1^T & 0 \end{bmatrix} x, \tag{3.4}$$

where A_1, E_1, E_2 are symmetric, A_1 is negative semidefinite, E_1 is positive semidefinite, E_2 is positive definite, and B_1 is an incidence matrix defining the terminals. The impedance modeling of RC circuits (i.e., in the absence of inductances) yields systems of DAE index 1 consisting of the upper left blocks of (3.4). If a system is voltage driven (admittance form), i.e., the inputs are terminal voltages provided by voltage sources and the outputs are terminal currents, it is possible to rewrite the system in impedance form [25]. Please note that RLC circuits of the form above are always passive. More details about this characteristic, about the system structure of MNA modeled RLC circuits, and about projection type reduction techniques for such current and voltage driven circuits are available, e.g., in [22]. The corresponding transfer function of system (3.4) is

$$G(s) = B^T (sE - A)^{-1} B. \tag{3.5}$$

So far, MOR methods, i.e., the approximation of the system's transfer function (3.3), concentrate on reducing the order n of the system to a smaller order n_r. Under the assumption that the dimensions of the input and output vectors are much smaller than the order of the system itself, i.e.,

$$m, p \ll n, \tag{3.6}$$

the important information is identified within the reduced model. Only the states containing this input-output information are preserved. A schematic overview of this conventional MOR approach is given in Fig. 3.1.

For the moment, we assume the reduced-order system $\hat{G}_r(s)$ to have a fixed order n_r. The goal is to find $\hat{G}_r(s) \approx G(s)$ such that $\|G(.) - \hat{G}_r(.)\|_*$ in minimized in some appropriate norm $\|\cdot\|_*$. This will yield a small output error $\|y(.) - \hat{y}(.)\|_*$ in frequency domain, and, ideally, after applying the inverse Laplace transform, also in time domain. The smaller n_r, the larger is the error, such that it is an additional challenge to find the smallest possible n_r for a given tolerance.

As already mentioned in Sect. 3.1, the assumption (3.6) is often violated due to new tasks in applications. The circuit simulation of power grids as part of an IC is one of the most common examples for these kinds of systems. Before we present different methods to handle MOR including, at least transitional, terminal reduction, we need a few more basic definitions.

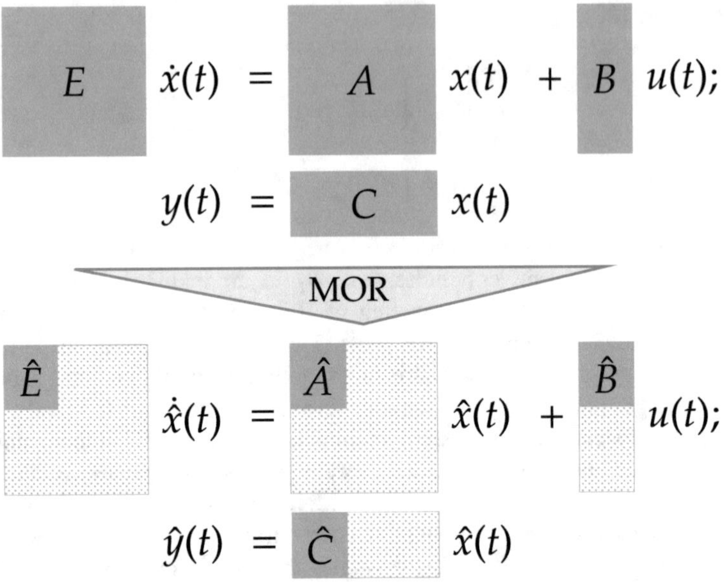

$$E \quad \dot{x}(t) \; = \; A \quad x(t) \; + \; B \; u(t);$$

$$y(t) \; = \; C \quad x(t)$$

MOR

$$\hat{E} \qquad \dot{\hat{x}}(t) \; = \; \hat{A} \qquad \hat{x}(t) \; + \; \hat{B} \quad u(t);$$

$$\hat{y}(t) \; = \; \hat{C} \quad \hat{x}(t)$$

Fig. 3.1 Schematic overview of conventional MOR

Definition 3.2.2 (The *i*-th Block Moment) We define the matrix $\mathbf{m_i^0} \in \mathbb{C}^{p \times m}$ as

$$\mathbf{m_i^0} = \begin{bmatrix} m_{i\,1,1}^0 & m_{i\,1,2}^0 & \cdots & m_{i\,1,m}^0 \\ m_{i\,2,1}^0 & m_{i\,2,2}^0 & \cdots & m_{i\,2,m}^0 \\ \vdots & \vdots & \ddots & \vdots \\ m_{i\,p,1}^0 & m_{i\,p,2}^0 & \cdots & m_{i\,p,m}^0 \end{bmatrix} = C(A^{-1}E)^i(-A)^{-1}B, \qquad i = 0, 1, \ldots$$

as the *i-th block moment of (3.3)*.

The matrices $\mathbf{m_i^0}$ are equal to the coefficients of the Taylor series expansion of (3.3) about $s_0 = 0$,

$$G(s) = C(sE - A)^{-1}B = C(I_n + (s - 0)\mathscr{K})^{-1}\mathscr{L} = \sum_{i=0}^{\infty} (-1)^i C \mathscr{K}^i \mathscr{L}(s - 0)^i,$$

with $\mathscr{K} := (-A)^{-1}E$ and $\mathscr{L} := (-A)^{-1}B$.

Definition 3.2.3 (The i-th s_0 Frequency-Shifted Block Moment) We define the matrix $\mathbf{m_i^{s_0}} \in \mathbb{C}^{p \times m}$ as

$$\mathbf{m_i^{s_0}} = \begin{bmatrix} m_{i\ 1,1}^{s_0} & m_{i\ 1,2}^{s_0} & \cdots & m_{i\ 1,m}^{s_0} \\ m_{i\ 2,1}^{s_0} & m_{i\ 2,2}^{s_0} & \cdots & m_{i\ 2,m}^{s_0} \\ \vdots & \vdots & \ddots & \vdots \\ m_{i\ p,1}^{s_0} & m_{i\ p,2}^{s_0} & \cdots & m_{i\ p,m}^{s_0} \end{bmatrix} = C(-(s_0 E - A)^{-1} E)^i (s_0 E - A)^{-1} B, \quad i = 0, 1, \ldots$$

as the i-th s_0 *frequency-shifted block moment* of (3.3).

The matrices $\mathbf{m_i^{s_0}}$ are equal to the coefficients of the Taylor series expansion of (3.3) about $s_0 \neq 0$

$$G(s) = C(sE - A)^{-1} B = C(I_n + (s - s_0)\mathcal{K})^{-1}\mathcal{L} \tag{3.7}$$

$$= \sum_{i=0}^{\infty} (-1)^i C \mathcal{K}^i \mathcal{L}(s - s_0)^i, \tag{3.8}$$

with $\mathcal{K} := (s_0 E - A)^{-1} E$ and $\mathcal{L} := (s_0 E - A)^{-1} B$.

The known moment matching methods [20, 21] make use of the fact that for $s \approx s_0$, the leading moments carry sufficient information about the dynamics of the system to approximate its input-output behavior. A detailed inspection of the block moments reveals that the j-th row, $j = 1, \ldots, p$, of $\mathbf{m_i^{s_0}}$, where m_i is one of the i-th (s_0 frequency-shifted) block moments, contains information how the output terminal j is influenced by all inputs. Analogously, the k-th column of $\mathbf{m_i^{s_0}}$, $k = 1, \ldots, m$, provides information about the impact of the input signal at terminal k to all outputs. Inspired by this observation, we define two moment matrices formed by κ different matrices $\mathbf{m_i^{s_0}}$.

Definition 3.2.4 (The Input Response Moment Matrix M_I) The matrix $M_I \in \mathbb{C}^{\kappa p \times m}$ composed of κ different block moment matrices m_i

$$M_I = \begin{bmatrix} \mathbf{m_0}^T, \mathbf{m_1}^T, \ldots, \mathbf{m_{\kappa-1}}^T \end{bmatrix}^T \tag{3.9}$$

is called *the input response moment matrix M_I of order κ* of (3.3).

Definition 3.2.5 (The Output Response Moment Matrix M_O) The matrix $M_O \in \mathbb{C}^{\kappa m \times p}$ composed of κ different block moment matrices m_i

$$M_O = [\mathbf{m_0}, \mathbf{m_1}, \ldots, \mathbf{m_{\kappa-1}}]^T \tag{3.10}$$

is called *the output response moment matrix M_O of order κ* of (3.3).

Remark 3.2.6 The choice of $\mathbf{m_i}$ to create M_I and M_O, i.e., whether to use $\mathbf{m_i^0}$ or $\mathbf{m_i^{s0}}$, and of which order κ, is free. However, the following facts give hints. The calculation of the moments requires computational effort of iterative character. In addition, lower order moments often contain basic information and the computation of higher order moments might be numerically unstable. Therefore, it is recommended to use all moments $\mathbf{m_i}$ up to a certain order κ. Frequency shifted block moments $\mathbf{m_i^{s0}}$ are even more expensive to compute. Consequently, making use of them is only recommended if a certain frequency is of special interest or there is a large approximation error at this frequency. The order κ of M_I and M_O does not need to be equal, but since the information resulting from the most expensive computational steps can be shared, it is obviously beneficial to use the same κ. In the symmetric case, only the computation of one of the response moment matrices is necessary. For more information. please see Sect. 3.4.3.

3.2.2   Benchmark Systems

In the following, we introduce two examples which accompany this chapter. If numerical experiments and results are shown they are with reference to one of the following systems.

3.2.2.1   A Test Circuit Example

The first numerical example was provided by the former Qimonda AG, Munich, Germany. It is a very simple parasitic RC test circuit called *RC549*, which is a linear subcircuit of a much larger nonlinear circuit. The model consists of one hundred and forty-one nodes, such that we get $n = 141$ generalized states in the corresponding descriptor system equations. Nearly half of these nodes, more precisely 49.65%, are terminals. As these terminals are the interconnects to the full circuit, it follows that $m = p = 70$.

Circuit *RC549* is a very useful test example because although half of the states are terminals, computations do not need a long time, such that a lot of tests can be performed in short time. Therefore, this test example was investigated beforehand also in [6, 7, 30].

3.2.2.2   Linear Subdomain for Non-linear Electro-Quasistatic Field Simulations

This example shows that the introduced methods are also applicable to problems which are not results of circuit simulation modeling. It also shows that the algorithms are applicable to large-scale systems. The simulation of high voltage models often includes nonlinear field grading materials, typically as thin layers. They allow higher voltage levels for the modeled devices. These materials have a highly nonlinear conductivity, leading to nonlinear resistive effects. A common way to describe such models is the electro-quasistatic (EQS) approximation. A standard finite element method (FEM) discretization results in a stiff system of ordinary differential equations (ODEs). To avoid the evaluation of all system equations during the integration, MOR is applied to a subdomain, e.g., proper orthogonal decomposition (POD), see [42]. Since most of the domain has constant material parameters (nonlinear field grading materials are used as thin layers), the decoupling of the system in a large linear part (upper left block) and a small conductive part (lower right block) including nonlinearities, such that

$$\begin{bmatrix} M_{11} & M_{12} \\ M_{21} & M_{22} \end{bmatrix} \begin{bmatrix} \dot{x}_1(t) \\ \dot{x}_2(t) \end{bmatrix} + \begin{bmatrix} K_{11} & K_{12} \\ K_{21} & K_{22} \end{bmatrix} \begin{bmatrix} x_1(t) \\ x_2(t) \end{bmatrix} = \begin{bmatrix} b_1(t) \\ b_2(t) \end{bmatrix},$$

is possible. For details, see [42]. The matrices M and K denote the discrete div-grad operators with respect to permittivity and conductivity. The state vector $x(t)$ contains the nodes potentials. The vector $b(t)$ contains boundary condition information, but only the components of $b(t)$ with real input are nonzero. Therefor, $b(t)$ can be seen as the vector of inputs mapped to the system by an incidence matrix, such that $b(t) = \hat{B}u(t)$, resulting in the system equations

$$\begin{bmatrix} M_{11} & M_{12} \\ M_{21} & M_{22} \end{bmatrix} \begin{bmatrix} \dot{x}_1(t) \\ \dot{x}_2(t) \end{bmatrix} + \begin{bmatrix} K_{11} & K_{12} \\ K_{21} & K_{22} \end{bmatrix} \begin{bmatrix} x_1(t) \\ x_2(t) \end{bmatrix} = \begin{bmatrix} \hat{B}_1 \\ \hat{B}_2 \end{bmatrix} \begin{bmatrix} u_1(t) \\ u_2(t) \end{bmatrix}.$$

Having a look at the second row, we get

$$M_{22}\dot{x}_2(t) + K_{22}x_2(t) = \hat{B}_2 u(t) - M_{21}\dot{x}_1(t) - K_{21}x_1(t), \tag{3.11}$$

which represents only the linear part, input information u and interconnect information to the conductive subdomain variables $\dot{x}_1(t)$ and $x_1(t)$. A reformulation of this equation leads to a system equivalent to system (3.1), i.e.,

$$\underbrace{M_{22}}_{:=E} \dot{x}_2(t) = \underbrace{-K_{22}}_{:=A} x_2(t) + \underbrace{\begin{bmatrix} \hat{B}_2 & -K_{21} & -M_{21} \end{bmatrix}}_{:=B} \begin{bmatrix} u(t) \\ x_1(t) \\ \dot{x}_1(t) \end{bmatrix}. \tag{3.12}$$

W.l.o.g., we assume $C = B^T$. The system, called eqs_lin in this work, is of order $n = 14{,}261$ with $m = p = 2943$ terminals (20.6%).

3.3 Terminal Reduction Approaches

In this section we will provide an overview about some existing methods which tackle the problem of MOR for systems with a high number of terminals (inputs/outputs). In general, we assume that the original system has too many input and/or output terminals to use conventional MOR approaches, so that the extra step of *terminal reduction* is necessary.

3.3.1 (E)SVDMOR

Having a closer look at (3.1), we recognize that the matrices C and B carry the input and output information. Trying to approximate (3.3) considering that (3.6) is violated motivates the idea to modify these input/output matrices. For this purpose, we try to find a projection of the $(p \times m)$-transfer function $G(s)$ onto a $(r_o \times r_i)$-transfer function $G_r(s)$ such that $r_o \ll p$ and $r_i \ll m$. To achieve this, we look for a decomposition of the transfer function such that

$$\widehat{G}(s) \approx V_C \underbrace{\left[W_C^T C(sE - A)^{-1} BV_B \right]}_{:=G_r(s)} W_B^T \tag{3.13}$$

with $V_C, W_C \in \mathbb{R}^{p \times r_o}$ and $V_B, W_B \in \mathbb{R}^{m \times r_i}$. Of course, the projector properties $W_C^T V_C = I_{r_o}$ and $W_B^T V_B = I_{r_i}$ should hold. The so-obtained internal transfer function $G_r(s)$ can be interpreted as a transfer function of a dynamical system with fewer virtual input and output terminals. This system can be further reduced with any method for model reduction of linear descriptor systems. Again, a schematic overview of this approach is shown in Fig. 3.2.

In what follows, we try to make use of the intimate correlations between the input and the output terminals. ESVDMOR [32, 33, 35, 47], an extended version of SVDMOR [16, 47], is a method which is based on the SVD of the input and the output response moment matrices M_I and M_O, see Definitions 3.2.4 and 3.2.5. Note that we assume the number of rows in each of both matrices to be larger than the number of columns. If this is not the case, the order κ has to be increased.

Applying the SVD, we get a low rank approximation of the form

$$M_I = U_I \Sigma_I V_I^* \approx U_{I_{r_i}} \Sigma_{I_{r_i}} V_{I_{r_i}}^*, \qquad M_O = U_O \Sigma_O V_O^* \approx U_{O_{r_o}} \Sigma_{O_{r_o}} V_{O_{r_o}}^*, \tag{3.14}$$

where

- for any matrix Z, Z^* is its conjugate transpose—in case $s_0 = 0$, the SVD is real and we can work with just the transpose Z^T, but for complex s_0, the moment matrices are in general complex, too;

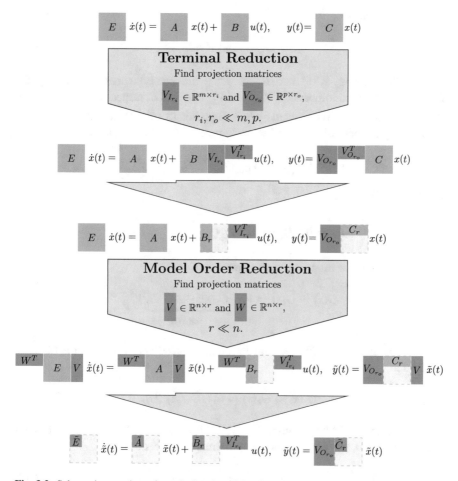

Fig. 3.2 Schematic overview of terminal and model order reduction

- $\Sigma_{I_{r_i}}$ is an $r_i \times r_i$ diagonal matrix and $\Sigma_{O_{r_o}}$ is an $r_o \times r_o$ diagonal matrix;
- $V_{I_{r_i}}$ and $V_{O_{r_o}}$ are $m \times r_i$ and $p \times r_o$ isometries (i.e., matrices having orthogonal/unitary columns) containing the dominant row subspaces of M_I and M_O, respectively;
- $U_{I_{r_i}}$ and $U_{O_{r_o}}$ are $\kappa p \times r_i$ and $\kappa m \times r_o$ isometries that are not used any further,
- r_i and r_o are in each case the numbers of significant singular values. At the same time they are the numbers of the reduced virtual input and output terminals.

For ease of notation, we furthermore assume that a real expansion point s_0 was chosen, so that the SVDs (3.14) are real and we can work with transposes in the following.

The important information about the input and output correlations is now contained in $V_{I_{r_i}}^T$ and $V_{O_{r_o}}^T$. Our goal is to decompose C and B. Combining this information leads to

$$C \approx V_{O_{r_o}} C_r \text{ and } B \approx B_r V_{I_{r_i}}^T. \tag{3.15}$$

The matrices $C_r \in \mathbb{R}^{r_o \times n}$ and $B_r \in \mathbb{R}^{n \times r_i}$ are consequences of applying the Moore-Penrose pseudoinverse of $V_{O_{r_o}}$ and $V_{I_{r_i}}^T$ to C and B, respectively. The Moore-Penrose pseudoinverse is denoted by $(\cdot)^+$. Equation (3.15), solved for C_r and B_r and modified by the definition of the Moore-Penrose pseudoinverse, leads to

$$C_r = V_{O_{r_o}}^+ C = (V_{O_{r_o}}^T V_{O_{r_o}})^{-1} V_{O_{r_o}}^T C = V_{O_{r_o}}^T C \tag{3.16}$$

and

$$B_r = B \left(V_{I_{r_i}}^T \right)^+ = B V_{I_{r_i}} (V_{I_{r_i}}^T V_{I_{r_i}})^{-1} = B V_{I_{r_i}}. \tag{3.17}$$

Hence, we get a new internal transfer function $G_r(s)$

$$G(s) \approx \widehat{G}(s) = V_{O_{r_o}} \underbrace{C_r(sE - A)^{-1} B_r}_{:=G_r(s)} V_{I_{r_i}}^T,$$

which is equivalent to (3.13). The terminal reduced transfer function $G_r(s)$ is reduced to

$$G_r(s) \approx \tilde{G}_r(s) = \tilde{C}_r(s\tilde{E} - \tilde{A})^{-1} \tilde{B}_r \tag{3.18}$$

by any conventional MOR method. The result is a very compact terminal and order reduced model $\tilde{G}_r(s)$. The complete approximation procedure is

$$G(s) \approx \widehat{G}(s) = V_{O_{r_o}} G_r(s) V_{I_{r_i}}^T \approx \widehat{G}_r(s) = V_{O_{r_o}} \tilde{G}_r(s) V_{I_{r_i}}^T. \tag{3.19}$$

Remark 3.3.1 SVDMOR can be considered as a special case of ESVDMOR setting (3.9) and (3.10) up with $\kappa = 1$ and $\mathbf{m}_0^0$. Furthermore, in [16] a recursive decomposition of the matrix-transfer function and an application of SVDMOR to each block, the so called RecMOR approach, is suggested.

Figures 3.3, 3.4, 3.5 and 3.6 show the reduction of *RC549* to 5 generalized states via one virtual terminal by ESVDMOR. The reduction step in (3.18) is performed by balanced truncation, in particular, using the implementation called "generalized square root method" [38]. In Fig. 3.3 it is clearly identifiable that one singular value of $M_I = m_0^0$, and therefore also of $M_O = \left(m_0^0 \right)^T$, dominates. Consequently, the system can be reduced to one virtual terminal. The calculation of frequency-shifted or higher order moments is more costly and in this example not necessary. For

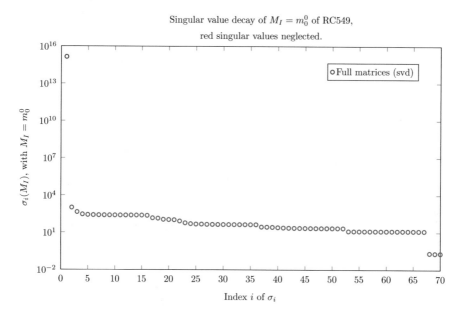

Fig. 3.3 Singular value decay of *RC549*, $M_I = m_0^0$, one virtual terminal

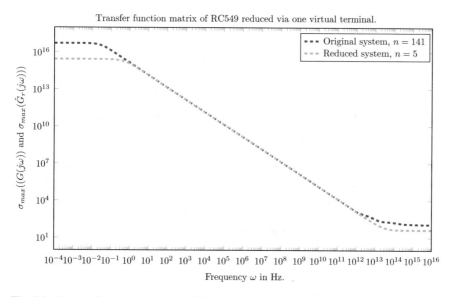

Fig. 3.4 Transfer function matrix of *RC549* reduced with ESVDMOR to 5 states via 1 virtual terminal

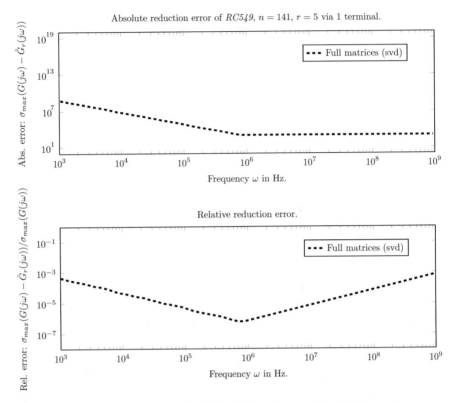

Fig. 3.5 Absolute and relative error of *RC549* reduced to 5 states via 1 virtual terminal

now, we follow the approach to explicitly calculate the matrices M_I and M_O and perform a full SVD in (3.14), see *Full matrices (svd)* in the legends. In contrast to the TermMerg approach, see the next subsection, the ESVDMOR approach does not allow to directly identify which state represents the dominant behavior of the terminals best. Since the spectral norm plot of the transfer function, see Fig. 3.4, and the magnitude plot of node one of the system, see Fig. 3.6, are quite similar, we can conclude that node one is representative for all terminals. This is expectable for all terminal nodes because of the one dominant singular value. The phase plot makes clear that the approximation within the frequency range of importance between 10^3 and 10^9 Hz is sufficient. Outside of this range, the phase approximation is not accurate. Figure 3.5 shows the overall approximation error of the reduced system. Although the absolute errors seems to be very large, the relative errors show that the approximation in spectral norm is acceptable, at least in the frequency range of interest.

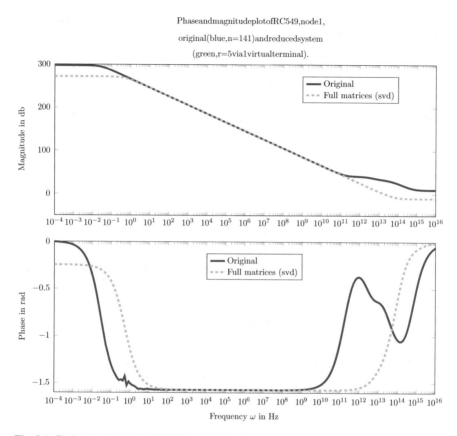

Fig. 3.6 Bode plot of node 1 of *RC549* reduced to 5 states via 1 virtual terminal

3.3.2   *TermMerg*

Based on a similar idea as ESVDMOR, the TermMerg approach was introduced in [34]. As can be suspected by its name, the approach merges terminals. If some terminals are similar in terms of performance metrics (timing, delays), those are identified and grouped as one terminal. TermMerg is, as well as ESVDMOR, based on the higher order moment responses. It additionally takes time delays into account while clustering the terminals. The input and output terminals are clustered separately. Terminals with similar timing responses are merged to one representative terminal. The number of clusters for the grouping is calculated by an SVD low rank approximation of the input and output response moment matrices M_I and M_O. The k-means clustering algorithm is used to group the column vectors of these matrices to different clusters. For each of them, one terminal is selected to represent all other terminals of this cluster. Adaptation of the input and output position matrices B and

C with respect to the representing vector of each cluster leads to a reduction in the number of terminals.

3.3.2.1 The k-Means Clustering Algorithm

The k-means clustering is a cluster analysis method to partition n observations into k clusters Cl_i, $i = 1, 2, \ldots, k$. Each observation belongs to the cluster with the nearest mean. We take a set of observations $(x_1, x_2, \ldots, x_n)$, with $x_j \in \mathbb{R}^\ell$ for $j = 1, \ldots, n$, for granted. The k-means clustering aims to partition the n observations such as to minimize the within-cluster sum of squares

$$S = \sum_{i=1}^{k} \sum_{x_j \in Cl_i} \|x_j - \mu_i\|^2. \tag{3.20}$$

The vector μ_i represents the geometric centroids of the data in Cl_i. The final result is a set of k disjoint subsets Cl_i containing N_i data points each and $n = \sum_{i=1}^{k} N_i$. One of the most commonly used k-means algorithms is the Lloyd algorithm [36] which gives the global minimum of S in (3.20). The three main steps of the algorithm are:

- Initialization: Set k initial means $\mu_1{}^1, \ldots, \mu_k{}^1$.
- Assignment: Each observation is assigned to the cluster with the closest mean,

$$Cl_i{}^t = (x_j : \|x_j - \mu_i{}^t\| \leq \|x_j - \mu_{i*}{}^t\|),$$

for $j = 1, \ldots, n$, $i* = 1, \ldots, k$, and t denotes the iteration number, and
- Update: The new center vectors of each cluster are calculated by,

$$\mu_i^{t+1} = \frac{1}{N_i^t} \sum_{x_j \in Cl_i{}^t} x_j.$$

The algorithm is repeated until there is no more change in the assignment step. The structure of this algorithm is responsible for the sensitivity to the initial choice of the number of clusters and their cluster centers. Fortunately, we calculate the number of clusters beforehand by using the truncated SVD in (3.14). Note that the number of reduced inputs r_i and r_o is equivalent to the number of input and output clusters, respectively.

Assume $x_j, j = 1, \ldots, m$, to be a column vector of M_I containing the moments w.r.t. to one input node (node j) and all output nodes such that

$$M_I = [x_1, x_2, \ldots, x_m].$$

Following the Lloyd algorithm, we pick k different vectors out of the x_j's as initial means, perform the steps described above, and save the resulting clusters $Cl_1, \ldots, Cl_{r_i}$ as well as the corresponding centers $\mu_1, \ldots, \mu_{r_i}$. Within TermMerg, we select one representative terminal for each cluster. For each Cl_i we find the element x_j for which the distance from μ_i is the minimum among all elements in Cl_i.

Example 3.3.2 We consider a very simple matrix A defined as follows:

$$A = \begin{bmatrix} 1 & 3 & 5 & 7 \\ 2 & 4 & 6 & 8 \end{bmatrix}.$$

Suppose we group the four columns into two clusters Cl_1 and Cl_2. We choose the initial centroids randomly as $\mu_1{}^1 = (3,4)^T$ and $\mu_2{}^1 = (5,6)^T$. We compute a distance matrix containing the squared Euclidean distances of each element (column index) to each cluster (row index). For the first element $a_1 = (1,2)^T$, the distance from the first centroid is given by $(1-3)^2 + (2-4)^2 = 8$. We construct the distance matrix as

$$D = \begin{bmatrix} 8 & 0 & 8 & 32 \\ 32 & 8 & 0 & 8 \end{bmatrix}.$$

The first set of clusters is $Cl_1{}^1 = \{a_1, a_2\}$ and $Cl_2{}^1\{a_3, a_4\}$. The new centroids are $\mu_1{}^2 = (2,3)^T$ and $\mu_2{}^2 = (6,7)^T$. We repeat the process, and the second set of clusters turns out to be the same as the first one. Consequently, the centroids stay the same. Since there is no further change in the centroids, the four columns can be clustered into $Cl_1{}^2$ and $Cl_2{}^2$.

3.3.2.2   The Reduction Step

Once we know the number of clusters for inputs and outputs, we perform k-means to cluster the terminals. We find the representative terminal for each cluster and accordingly replace columns of the input position matrix B and the rows of the output position matrix C. We include only the representative terminals.

3.3.3   SparseRC

In this subsection we comment on an approach based on a completely different idea from the one used in the approaches we introduced up to now. Indeed, SparseRC, which is introduced in [25, 26], also performs MOR although the number of terminals may be very large. A specialty is that the approach is considerably more user-oriented. Its application is the reduction of parasitic RC circuits in simulation, e.g. power grids. In [26], the following five basic objectives for successful model reduction are mentioned:

- the approximation error $\|G(.) - \hat{G}(.)\|$ is small in an appropriate norm,
- stability and passivity are preserved,
- an efficient computation is possible,
- for reconnection, the incidence input/output matrix B is preserved in some sense, i.e., $\hat{B}$ is a sub-matrix of B, and
- the reduced system matrices $\hat{E}$ and $\hat{A}$ are sparse.

While the first three items can be nearly taken for granted in MOR, the last two items are user-oriented specifications which are hard to meet. Even if the last item may be satisfied by some approaches, the next-to-last item is in general not satisfied as in most conventional MOR methods, the reduced model is not sparse. The SparseRC approach promises to fulfill these five items. The basic idea is to apply extended moment matching projection (EMMP), see next section, to disjoint parts of the original system.

3.3.3.1 MOR via Graph Partitioning and EMMP

Following [26], the EMMP (derived from an algorithm based on pole analysis via congruence transformations called PACT [28] and from sparse implicit projection (SIP), see [48]) is a moment matching type reduction approach suitable for multi-terminal systems with only a few terminals. Within SparseRC, this projection method is applied to subsystems connected via so called separator nodes. Each subsystem has just a few terminals and the connection of all reduced subsystems leads to a reduced-order version of the original system.

Assume such a subsystem is given as an LTI system obtained from modeling an RC circuit. In this situation, system (3.4), re-arranged and in frequency domain, can be written as

$$(sE - A)x(s) = Bu(s), \tag{3.21}$$

when ignoring the output term. Now we do a simple block separation into selected nodes x_s (terminals and separator nodes) and internal nodes x_i. This leads to

$$\left(s \begin{bmatrix} E_i & E_c \\ E_c^T & E_s \end{bmatrix} - \begin{bmatrix} A_i & A_c \\ A_c^T & A_s \end{bmatrix} \right) \begin{bmatrix} x_i \\ x_s \end{bmatrix} = \begin{bmatrix} 0 \\ B_s \end{bmatrix} u,$$

where $A_i, E_i \in \mathbb{R}^{(n-s)\times(n-s)}$, $B_s \in \mathbb{R}^{s\times s}$ with $s \geq m$ being the number of selected nodes, i.e. the number of terminals m plus the separator nodes which are preserved because of the coupling within SparseRC. Applying a congruence transform with

$W := -A_i^{-1} A_c$ and

$$X = \begin{bmatrix} I_{n-s} & W \\ 0 & I_s \end{bmatrix}, \quad \bar{x} = X^T x$$

such that (A, E, B) are mapped to $(X^T A X, X^T E X, X^T B)$, yields

$$\left(s \begin{bmatrix} E_i & \bar{E}_c \\ \bar{E}_c^T & \bar{E}_s \end{bmatrix} - \begin{bmatrix} A_i & 0 \\ 0 & \bar{A}_s \end{bmatrix} \right) \begin{bmatrix} x_i \\ \bar{x}_s \end{bmatrix} = \begin{bmatrix} 0 \\ B_s \end{bmatrix} u, \tag{3.22}$$

with

$$\bar{E}_c = E_c + E_i W,$$

$$\bar{E}_s = E_s + W^T E_i W + W^T E_c + E_c^T W, \tag{3.23}$$

$$\bar{A}_s = A_s + A_c^T W. \tag{3.24}$$

Solving the first row in (3.22) for x_i and plugging the result into the equation of the second row of (3.22) leads to

$$(s\bar{E}_s - \bar{A}_s)\bar{x}_s - s^2 \bar{E}^T (sE_i - A_i)^{-1} \bar{E}_c \bar{x}_s = B_s u.$$

Following [26] further on, the first two moments of system (3.21) at $s = 0$ are given by $\bar{E}_s$ and $\bar{A}_s$ in (3.23) and (3.24). Consequently, a reduced-order model $\hat{A}, \hat{E} \in \mathbb{R}^{k \times k}$, $k \geq s$, which preserves the first two admittance moments of (3.21) can be computed via a moment matching projection $V = \begin{bmatrix} W^T & I_s \end{bmatrix}^T$ given by

$$\hat{A} = V^T A V, \qquad \hat{B} = V^T B,$$

$$\hat{E} = V^T E V, \quad \text{and} \quad \hat{x} = V^T x = \bar{x}_s.$$

The question remains how to partition the original RC system with many terminals in an appropriate way such that we can apply EMMP to the subsystems. There are two goals during this process, which depend on each other. One is to find a partitioning and the second is to avoid fill-in such that the reduced model is still sparse. Usually, fill-in minimizing matrix reordering techniques similar to those used to compute an efficient Cholesky factorization can identify nodes causing fill-in. SparseRC avoids these techniques due to circuit size or topology. Instead, graph decomposition based on the non-zero pattern (nzp) of $E - A$ is used due to the analogy between circuit models and graphs. The fill-creating nodes, i.e., the separation nodes, are identified and the circuit matrices are reordered in a so-called bordered block diagonal (BBD) form [49]. In this form, the individual blocks form the subsystems to be reduced and the border blocks contain the separator nodes, which are preserved during the EMMP. Both goals are achieved and the five objectives for MOR above are fulfilled. For a detailed discussion, e.g., about the matrix properties such as the singularity of A_i or implementation details, see [25, 26]. The approach is generalized to RLC networks in [25].

3.3.4 MOR for Many Terminals via Interpolation

This section deals with the modeling of RLC systems of the form (3.1) by means of S-parameters (scattering parameters), i.e., measurements of frequency domain response data [29]. A typical application would be a black box problem. Given is an electrical device, but the knowledge about the circuit inside the device is unidentified. With the help of, e.g., a Vector Network Analyzer (VNA), it is possible to measure amplitude and phase properties. We get P samples at different frequencies $f_j, j = 1, \ldots, P$, of the device with m inputs and p outputs as

$$
\mathbf{S}^{(j)} := \begin{bmatrix} S_{11}^{(j)} & \cdots & S_{1m}^{(j)} \\ \vdots & \vdots & \vdots \\ S_{p1}^{(j)} & \cdots & S_{pm}^{(j)} \end{bmatrix}. \tag{3.25}
$$

The goal is to find the associated transfer function $G(s)$ such that its values at $s = i\omega$ with $\omega = 2\pi f_j$ are close to the scattering matrix. If $\|G(i\,2\pi f_j) - \mathbf{S}^{(j)}\|$ is small in some norm, the computed representation matrices belong to an accurate model. The approach uses tangential interpolation and the Loewner matrix concept to find a model, even a reduced-order model, of this multiport device.

3.3.4.1 Tangential Interpolation and the Loewner Concept

Tangential interpolation is a form of the standard interpolation problem along a tangential direction. Consider k different sampling points $\lambda_i \in \mathbb{C}$, the tangential directions $r_i \in \mathbb{C}^{m \times 1}$, and the tangential data $w_i \in \mathbb{C}^{p \times 1}$, such that

$$
\Lambda = \{\lambda_1, \ldots, \lambda_k\} \subset \mathbb{C}^k,
$$
$$
R = [r_1, \ldots, r_k] \in \mathbb{C}^{m \times k}, \text{ and}
$$
$$
W = [w_1, \ldots, w_k] \in \mathbb{C}^{p \times k}
$$

is called the right interpolation data. Analogously, we define the left interpolation data $\theta_j \in \mathbb{C}$, $l_j \in \mathbb{C}^{1 \times p}$ and $v_j \in \mathbb{C}^{1 \times m}$ at another h sampling points as

$$\Theta = \{\theta_1, \ldots, \theta_k\} \subset \mathbb{C}^h,$$

$$L = \left[l_1^T, \ldots, l_h^T \right]^T \in \mathbb{C}^{h \times p}, \text{ and}$$

$$V = \left[v_1^T, \ldots, v_h^T \right]^T \in \mathbb{C}^{h \times m}.$$

The rational interpolation problem aims at finding a set of system matrices, such that the right and left interpolation constrains

$$G(\lambda_i) r_i = w_i, \quad i = 1, \ldots, k, \tag{3.26}$$

$$\text{and} \quad l_j G(\theta_j) = v_j, \quad j = 1, \ldots, h, \tag{3.27}$$

are fulfilled by the associated transfer function $G(s)$.

Definition 3.3.3 (The Loewner and the Shifted Loewner Matrix) Given are a set of $P = k + h$ sampling points $\{\lambda_1, \ldots, \lambda_k, \theta_1, \ldots, \theta_h\}$ and the values of a rational matrix function $G(\lambda_i)$ and $G(\theta_j)$ evaluated at these points. By defining tangential directions r_i and l_j we are able to compute the tangential data w_i and v_j. The matrix

$$\mathbb{L} = \begin{bmatrix} \frac{v_1 r_1 - l_1 w_1}{\theta_1 - \lambda_1} & \cdots & \frac{v_1 r_k - l_1 w_k}{\theta_1 - \lambda_k} \\ \vdots & \vdots & \vdots \\ \frac{v_h r_1 - l_h w_1}{\theta_h - \lambda_1} & \cdots & \frac{v_h r_k - l_h w_k}{\theta_h - \lambda_k} \end{bmatrix}$$

is called the *Loewner matrix*. The *shifted Loewner matrix* is defined as

$$\sigma \mathbb{L} = \begin{bmatrix} \frac{\theta_1 v_1 r_1 - \lambda_1 l_1 w_1}{\theta_1 - \lambda_1} & \cdots & \frac{\theta_1 v_1 r_k - \lambda_k l_1 w_k}{\theta_1 - \lambda_k} \\ \vdots & \vdots & \vdots \\ \frac{\theta_h v_h r_1 - \lambda_1 l_h w_1}{\theta_h - \lambda_1} & \cdots & \frac{\theta_h v_h r_k - \lambda_k l_h w_k}{\theta_h - \lambda_k} \end{bmatrix}.$$

By means of the matrices of Definition 3.3.3 and their properties, e.g., they satisfy certain Sylvester equations and turn out to reveal some kind of tangential controllability and observability Gramians, we can solve the modeling problem for the given data.

Lemma 3.3.4 *Assume $k = h$ (half of P, if impossible ignore one sampling point) and $\det(z\mathbb{L} - \sigma\mathbb{L}) \neq 0$ for all $z \in \{\lambda_i\} \cup \{\theta_j\}$. Then $E = -\mathbb{L}$, $A = -\sigma\mathbb{L}$, $B = V$,*

$C = W$, and $D = 0$ is a minimal realization[1] of the system whose transfer function

$$G(s) = W(\sigma\mathbb{L} - s\mathbb{L})^{-1}V$$

interpolates the given data such that (3.26) and (3.27) are satisfied.

Note that we assume exact data and the needed number of samples. For a proof and more details we again point to [29]. There, it is also explained how to reveal the order of the underlying system, how to handle different scenarios of measurements, and how to get reduced-order models. This kind of approach can be extended to parametric systems, see [3].

Another important information is how to get the tangential data from (3.25). Note that, like in the scattering matrix, for real passive RC circuits the number of inputs is equal to the number of outputs $m = p$. In [29], it is proposed to define r_i and l_j as rows and columns of the identity matrix I_m to be linearly independent. Consequently, for $j = 1, \ldots, k$, the right interpolation data can be constructed as

$$\lambda_j = i\omega_j, \quad r_j = e_q, \quad \text{and} \quad w_j = S^{(j)}r_j,$$

where $e_q \in \mathbb{R}^{m \times 1}$ (an m-dimensional unit vector with entry 1 in its q-th position) with $j \equiv q \bmod m$. If $q = 0$ we set $q = m$, i.e., if the division of k and m is without residual, we take the last row of I_m as r_j. Analogously, for $j = 1, \ldots, h$ we get the left interpolation data as

$$\theta_j = -i\omega_j, \quad l_j = r_j^T, \quad \text{and} \quad v_j = l_j\overline{S^{(j)}}.$$

3.4 ESVDMOR in Detail

3.4.1 Stability, Passivity, Reciprocity

The preservation of properties is an important topic in MOR. Besides the essential information encoded in the equations of the original system (3.1), e.g., resistance values or capacities, other information, e.g., structural characteristics (block structure, sparsity) or physical qualities, can be of interest. The preservation of such properties is of prime importance in a lot of applications. Typically in circuit simulation, three basic properties to preserve are stability, passivity, and reciprocity; see also Sect. 2.2.3. For this reason, in the following section we give some basic definitions and lemmas to establish facts on the preservation of these three properties in reduced-order models generated by ESVDMOR [7].

---

[1]Loosely speaking, a *minimal realization* of a descriptor system is a set of matrices (A, B, C, E) of minimal order yielding the transfer function $G(s)$ of the system.

3.4.1.1 Stability

In the analysis of dynamical systems it is important to study the dynamics if the considered time horizon goes to infinity. The reduced system is intended to show the same behavior as the original one. The following basic definition and lemma can be found, e.g., in [10, 14].

Definition 3.4.1 (Asymptotic Stability, c-Stability)   The descriptor system (3.1) is *asymptotically stable* if all solutions $x(t)$ of $E\dot{x}(t) - Ax(t) = 0$ converge to zero at infinity, i.e.,

$$\lim_{t \to \infty} x(t) = 0.$$

The matrix pencil $\lambda E - A$, $\lambda \in \mathbb{C}$, is called *c-stable* if it is regular and the corresponding system (3.1) is asymptotically stable.

Lemma 3.4.2 *Consider a descriptor system (3.1) with a regular matrix pencil $\lambda E - A$. The following statements are equivalent:*

1. *System (3.1) is asymptotically stable.*
2. *The finite eigenvalues of the pencil $\lambda E - A$, $\lambda_i \in \Lambda(A, E)$, lie in the open left complex half-plane, i.e., $\lambda_i \in \mathbb{C}_- = \{\lambda \in \mathbb{C} |\, \mathrm{Re}(\lambda) < 0\}$.*
3. *Consider the spectral projections P_r and P_l, respectively, onto the right and the left deflating subspace of the pencil corresponding to the finite eigenvalues. The projected generalized continuous-time Lyapunov equation*

$$E^T XA + A^T XE + P_r^T QP_r = 0, \qquad X = P_l^T XP_l$$

has a unique Hermitian, positive semidefinite (psd) solution X for every Hermitian, psd right hand side matrix Q.

Remark 3.4.3 The infinite eigenvalues of the pencil do not affect the behavior of the homogeneous system in Definition 3.4.1. Consequently they do not affect stability. A very useful observation of Lemma 3.4.2 is the connection between the stability of the system and the finite eigenvalues of the pencil. We use this fact to prove the following theorem.

Theorem 3.4.4 *Assume the descriptor system (3.1) with its transfer function (3.3) to be asymptotically stable. The ESVDMOR reduced-order system corresponding to $\hat{G}_r(s)$ of (3.19) is asymptotically stable if the inner reduction to $\tilde{G}_r(s)$ of (3.18) is stability preserving.*

Proof According to Lemma 3.4.2, we know that in the continuous-time case stability is preserved if the pencil of the reduced system is c-stable. Besides regularity this means $\mathrm{Re}(\tilde{\lambda}) < 0$ for all $\tilde{\lambda} \in \Lambda(\tilde{A}, \tilde{E})$ (the spectrum of the matrix pencil $\lambda \tilde{E} - \tilde{A}$). Remembering that the original system $G(s)$ is asymptotically stable, i.e., $(\lambda E - A)$ is c-stable, the following implication is obvious:

The inner reduction $G_r(s) \approx \tilde{G}_r(s)$ in (3.18) is stability preserving
$\implies (\lambda \tilde{E} - \tilde{A})$ is c-stable
$\implies$ the reduction $G(s) \approx \hat{G}_r(s)$ in (3.19) is stability preserving.

Remark 3.4.5 Many established MOR methods for regular descriptor systems are stability preserving and can be applied along the lines of Theorem 3.4.4. Due to their computable error bounds we prefer the use of balanced truncation based approaches, see [5, 10], and in particular the specialized versions described in Chap. 2 dedicated to circuit equations.

3.4.1.2   Passivity

One of the most common definitions of passivity of a circuit is the property that its elements consume (or at least do not produce) energy. Thus, if the original system is passive, also the reduced model should be passive in order to remain physically meaningful. In other words, passivity preservation is important for stable, accurate, and interpretable simulation results, see also Chap. 2. A more mathematical definition of passivity is the following one, taken from [40].

Definition 3.4.6 (Passivity)   A descriptor system (3.1) is called *passive* (or *input-output-passive*) if $m = p$ and the system output $y : \mathbb{R} \to \mathbb{R}^m$ satisfies

$$\int_{t_0}^{T} u(t)^T y(t) dt \geq 0, \tag{3.28}$$

for all possible inputs $u : \mathbb{R} \to \mathbb{R}^m$, where $x(t_0) = 0$, the functions u and y are square integrable, and $[0 \leq t_0; T] \subseteq \mathbb{R}$ is the time period of interest.

Since for linear systems a shifting of the time horizon is not a problem, we can assume $t_0 = 0$. Passivity, such as in the previous definition, is hard to show. Therefore, we use another concept, discussed, e.g., in [20], for constructive passivity testing.

Definition 3.4.7 (Positive Real Transfer Function)   The transfer function (3.3) is *positive real* iff the following three assumptions hold:

(i) $G(s)$ has no poles in $\mathbb{C}_+ = \{s \in \mathbb{C}\,|\, \text{Re}(s) > 0\}$, and additionally there are no multiple poles of $G(s)$ on the imaginary axis $i\mathbb{R}$,
(ii) $G(\bar{s}) = \overline{G(s)}$ for all $s \in \mathbb{C}$,
(iii) $\text{Re}(x^* G(s) x) \geq 0$ for all $s \in \mathbb{C}_+$ and $x \in \mathbb{C}^m$, i.e. $G(s)^* + G(s) \geq 0$ for all $s \in \mathbb{C}_+$.

To show some results about passivity preservation of ESVDMOR, we need to know the connection between passivity and positive realness.

Lemma 3.4.8 *A descriptor system* (3.1) *is passive if, and only if, its transfer function* (3.3) *is positive real.*

Remark 3.4.9 The original proof can be found in [1]. In [1, 37, 41] another proof of this lemma shows the equivalence of passivity and the bounded realness of the scattering matrix function, which is nothing else than a Moebius transformation of $G(s)$. Subsequently, the equivalence of the bounded realness of this scattering matrix function and the positive realness of the transfer function is shown. Hence the lemma is proven.

As already mentioned in Definition 3.4.6, for passive systems we assume that the number of inputs is equal to the number of outputs: $m = p$. We furthermore assume (3.4) and (3.5). As before, the matrix pencil $\lambda E - A$ is assumed to be regular.

Theorem 3.4.10 *Consider a passive system of the form* (3.4) *with its transfer function* (3.5). *The ESVDMOR reduced-order system corresponding to $\widehat{G}_r(s)$ of* (3.19) *is passive if the inner reduction to $\widetilde{G}_r(s)$ of* (3.18) *is passivity preserving.*

Proof Following Lemma 3.4.8, we show that $\widehat{G}_r(s)$ in (3.19) is positive real. Hence, we show that the reduced system is passive. The i-th s_0 frequency-shifted block moments of (3.5) are

$$\mathbf{m}_i^{s_0} = B^T(-(s_0E - A)^{-1}E)^i(s_0E - A)^{-1}B,$$

with $\det(s_0E - A) \neq 0$. Following the technique used in [21], we define $J = \begin{bmatrix} I & 0 \\ 0 & -I \end{bmatrix}$ with appropriate block structure. The properties of A_1, E_1, and E_2 as well as the fact that $J = J^T$ and $J^2 = I$ lead to the following rules:

R1: $J = J^{-1}$,
R2: $JE = EJ$, hence $JEJ = E$,
R3: $(s_0E - A)^T = s_0E - JAJ = s_0JEJ - JAJ$, hence $(s_0E - A)^{-T} = (s_0JEJ - JAJ)^{-1} = J^{-1}(s_0E - A)^{-1}J^{-1} = J(s_0E - A)^{-1}J$,
R4: $B = JB$, and
R5: for every matrix X, Y and $i \in \mathbb{N}$, $(-X^{-1}Y)^i = X^{-1}(Y(-X)^{-1})^iX$ holds.

A straightforward calculation employing these rules shows that

$$
\begin{aligned}
\mathbf{m}_i^{s_0\,T} &= (B^T(-(s_0E - A)^{-1}E)^i(s_0E - A)^{-1}B)^T \\
&= B^T(s_0E - A)^{-T}\{(\underbrace{-(s_0E - A)^{-1}}_{-X^{-1}}\,\underbrace{E}_{Y})^i\}^T B \\
&\overset{(R5)}{=} B^T(s_0E - A)^{-T}\{(s_0E - A)^{-1}(E(-(s_0E - A)^{-1}))^i(s_0E - A)\}^T B \\
&= B^T(s_0E - A)^{-T}(s_0E - A)^T\{(E(-(s_0E - A)^{-1}))^i\}^T(s_0E - A)^{-T}B \\
&= B^T\{(E(-(s_0E - A)^{-1}))^i\}^T(s_0E - A)^{-T}B
\end{aligned}
$$

$$\overset{(E=E^T)}{=} \quad B^T((-(s_0E-A))^{-T}E)^i(s_0E-A)^{-T}B$$

$$\overset{(R3)}{=} \quad B^T(-J(s_0E-A)^{-1}JE)^i(J(s_0E-A)^{-1}J)B$$

$$\overset{(R5)}{=} \quad B^TJ(s_0E-A)^{-1}(JE(-J(s_0E-A)^{-1}))^i(s_0E-A)J(J(s_0E-A)^{-1}J)B$$

$$\overset{(R1,R2)}{=} \quad B^TJ(s_0E-A)^{-1}(E(-(s_0E-A)^{-1}))^i(s_0E-A)(s_0E-A)^{-1}JB$$

$$= \quad B^TJ(s_0E-A)^{-1}(E(-(s_0E-A)^{-1}))^iJB$$

$$\overset{(R4)}{=} \quad B^T(s_0E-A)^{-1}(E(-(s_0E-A)^{-1}))^iB$$

$$\overset{(R5\,\text{backwards})}{=} \quad B^T(-(s_0E-A)^{-1}E)^i(s_0E-A)^{-1}B$$

$$= \quad \mathbf{m}_i^{s_0}.$$

By means of (3.14) it follows from (3.9) and (3.10) that $M_I = M_O$, such that $V_{I_{r_i}}^T = V_{O_{r_o}}^T = V_r^T$. Hence

$$\widehat{G}(s) = V_r B_r^T(sE-A)^{-1}B_r V_r^T, \tag{3.29}$$

with B_r analogous to (3.17). If the MOR method used in (3.18) leads to a positive real transfer function, passivity is preserved.

In Definition 3.4.7 and Lemma 3.4.8 it is shown that the system (3.4) is passive if the Hermitian part of the transfer function along the imaginary axis is positive semidefinite, i. e., $G_H = \frac{1}{2}(G(j\omega) + G(j\omega)) \geq 0$. As an illustration, in Fig. 3.7 we show the smallest nonzero eigenvalue of the Hermitian part of the original transfer function $G_H(j\omega)$ and of the terminal, not yet order reduced transfer function $\widehat{G}_H(j\omega)$. In every case the systems are positive semidefinite. Figure 3.8 shows the relative difference of these smallest eigenvalues to those of the original system. As we only add zero eigenvalues in (3.19) by multiplying $V_{O_{r_o}} = V_r$ and $V_{I_{r_i}}^T = V_r^T$ from the left and the right to the positive real transfer function $\tilde{G}_r(s)$, positive semidefiniteness is preserved.

As inner reduction method, again the passivity-preserving balanced truncation approaches described in Chap. 2 can be employed.

3.4.1.3 Reciprocity

Another important property of MOR methods is reciprocity preservation, which is a requirement for the synthesis of the reduced-order model as a circuit. We assume the setting given in (3.4). An appropriate definition can be found, e.g., in [40].

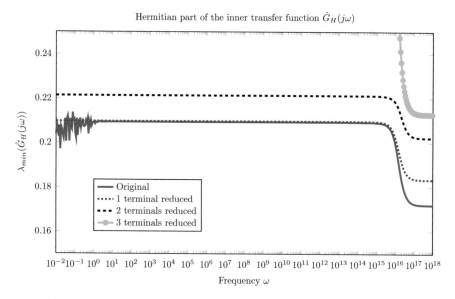

Fig. 3.7 Smallest eigenvalues of the Hermitian part of the transfer function of *RC549*

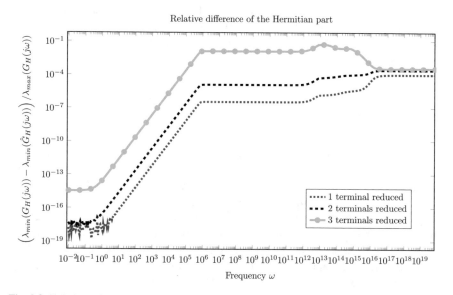

Fig. 3.8 Relative difference correlated to the largest eigenvalue of the original transfer function

Definition 3.4.11 A transfer function (3.3) is *reciprocal* if there exist $m_1, m_2 \in \mathbb{N}$ with $m_1 + m_2 = m$, such that for $\Sigma_e = \text{diag}(I_{m_1}, -I_{m_2})$ and all $s \in \mathbb{C}$ being not a pole of $G(s)$, it holds

$$G(s)\Sigma_e = \Sigma_e G^T(s).$$

The matrix Σ_e is called *external signature* of the system. A descriptor system is called *reciprocal* if its transfer function is reciprocal.

As a consequence, a transfer function of a reciprocal system is a matrix of the form

$$G(s) = \begin{bmatrix} G_{11}(s) & G_{12}(s) \\ -G_{12}^T(s) & G_{22}(s) \end{bmatrix}, \tag{3.30}$$

where $G_{11}(s) = G_{11}^T(s) \in \mathbb{R}^{m_1, m_1}$ and $G_{22}(s) = G_{22}^T(s) \in \mathbb{R}^{m_2, m_2}$. That is, $G(s)\begin{bmatrix} 0 & I \\ -I & 0 \end{bmatrix}$ is a Hamiltonian matrix.

Theorem 3.4.12 *Consider a reciprocal system of the form (3.4). The ESVDMOR reduced-order system corresponding to $\hat{G}_r(s)$ of (3.19) is reciprocal if the inner reduction to $\tilde{G}_r(s)$ of (3.18) is reciprocity preserving.*

Proof Due to the reciprocity of the original system, the corresponding transfer function (3.5) has the structure given in (3.30). Equation (3.29) shows that none of the steps in ESVDMOR destroy this symmetric structure if (3.18) preserves reciprocity.

3.4.2 Error Analysis

Knowledge about the approximation error is motivated by the desire to meet the (industrial) requirements placed on the reduced-order model and, at the same time, to get a reduced model, which is as small as possible and as good as necessary. Because approximation errors are, of course, caused at different steps of the algorithm, an analysis of these particular errors is an important basis to estimate the total approximation error. Most of the particular single errors are well studied. In the following section, we describe the results about a priori error estimation of the ESVDMOR approximation in [8]. Based on the known approximation errors, e.g. those of truncated SVD or balanced truncation, we combine all errors, firstly with many assumptions and later for more general models, to derive a global error bound of the ESVDMOR approach.

Since the difference between the original transfer function and the one given by the reduced-order model is still a matrix-valued function of s, matrix norms are needed to define the kind of error measurement. In the following, the two most important norms regarding MOR, the spectral norm and the $\mathscr{H}_\infty$-norm, are defined.

Definition 3.4.13 (Spectral Norm) The *spectral norm* of a matrix $H \in \mathbb{C}^{k \times \ell}$ is induced by the Euclidean vector norm and defined as

$$||H||_2 = \sqrt{\lambda_{\max}(H^*H)} = \sigma_{\max}(H),$$

where H^* denotes the conjugate transpose of H, $\lambda_{\max}(\cdot)$ the largest eigenvalue, and $\sigma_{\max}(\cdot)$ the largest singular value.

The second useful and important norm is based on the Hardy Space theory, see, e.g., [18].

Definition 3.4.14 ($\mathscr{H}_\infty$-Norm) The $\mathscr{H}_\infty$-*norm* of an asymptotically stable transfer function (3.3) is defined as

$$||G||_{\mathscr{H}_\infty} = \sup_{s \in \mathbb{C}_+} \sigma_{\max}(G(s)) = \sup_{s \in \mathbb{C}_+} ||G(s)||_2. \tag{3.31}$$

Due to the maximum modulus theorem, we can express (3.31) as $||G||_{\mathscr{H}_\infty} = \sup_{\omega \in \mathbb{R}} \sigma_{\max}(G(i\omega))$.

For unstable systems, the $\mathscr{H}_\infty$-norm is not defined. In this case, $G(s)$ is not holomorphic in the open right half plane, i.e., it is not an element of the Hardy space $\mathscr{H}_\infty$ of interest.

3.4.2.1 Particular Error Bounds

Equation (3.14) describes a truncated SVD. We know the error caused by an SVD approximation, e.g. of M_I, is given by

$$e_{M_I} = \left\| M_I(\kappa) - U_{I_{r_i}} \Sigma_{r_i}^I V_{I_{r_i}}^T \right\|_2 = \sigma_{r_i+1}^I,$$

where

$$\Sigma^I = \mathrm{diag}(\sigma_1^I, \ldots, \sigma_{r_i}^I, \sigma_{r_i+1}^I, \ldots, \sigma_{min}^I \geq 0), \quad \Sigma_{r_i}^I = \mathrm{diag}(\sigma_1^I, \ldots, \sigma_{r_i}^I),$$

and $\sigma_j^I \geq \sigma_{j+1}^I$ for $j = 1, \ldots, m_{in} - 1$. The same applies to M_O. The notation $M_I(\kappa)$ expresses the dependency on the number κ of used block moments m_i.

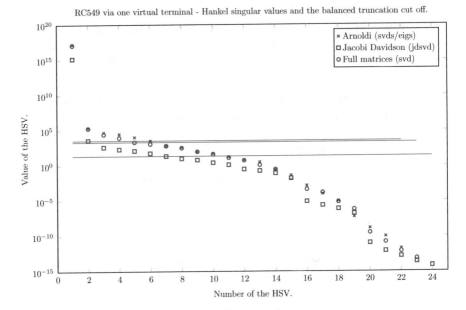

Fig. 3.9 Hankel singular value decay of *RC549* via one virtual terminal. The *red lines* show the BT error bounds, computed by the truncated HSVs for the three different methods

Provided the use of a suitable MOR method which provides an error bound, e.g., the balanced truncation (BT) methods for RC circuits as discussed in Chap. 2, we know the error caused in (3.18). BT methods perform a reduction based on the truncation of the so-called proper Hankel singular values (HSVs) $\hat{\sigma}_1, \ldots, \hat{\sigma}_q$ of the system (where $q \leq n$), see Sect. 2.3.2 for details. The error for balanced truncation using any of the variants discussed in the previous chapter is bounded by

$$\left\| G_r - \tilde{G}_r \right\|_{\mathscr{H}_\infty} \leq const. \cdot \sum_{k=n_r+1}^{q} \hat{\sigma}_k = \delta, \qquad (3.32)$$

in case we reduce the proper part of the system to order n_r.

Figure 3.9 shows the HSVs of *RC549*, for which the terminal reduced original system $G_r(s)$ is a single-input single-output (SISO) system, computed several methods (the Arnoldi and Jacobi-Davidson based methods are explained in the following section). We see that the decay of the HSVs is fairly slow. Additionally, the beforehand applied terminal reduction corrupts the computation, which explains the differences according to the decomposition approaches. Nevertheless, the strong decay of the HSVs and a relative tolerance of $\mathscr{O}(10^{-15})$ leads to only 5 (in JDSVD case to 6) reduced generalized states. Similarly, Fig. 3.10 shows the HSVs of *eqs_lin* with 573, respectively 520, virtual terminals. It is remarkable that the cut off in this

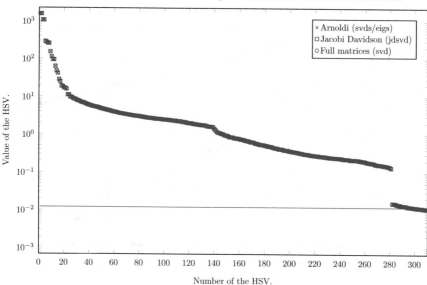

Fig. 3.10 Hankel singular value decay of *eqs_lin* via 573 (Arnoldi and full SVD) and 520 (JDSVD) virtual terminals

case was not induced by reaching the given relative tolerance but by accomplishing the predefined maximum order of the reduced system $n_r = 300$. As a result, we may expect a relative error of $\mathcal{O}(10^{-5})$, which is confirmed in Fig. 3.11.

3.4.2.2   Total ESVDMOR Error Bound

The error analysis discussed here follows in large parts [8]. Due to (3.19) and the triangle inequality, the total ESVDMOR error in the spectral norm on the imaginary axis can locally be expressed as

$$e_{tot} = \left\| G(i\omega) - \hat{G}_r(i\omega) \right\|_2 \leq \underbrace{\left\| G(i\omega) - \hat{G}(i\omega) \right\|_2}_{=e_{out}} + \underbrace{\left\| \hat{G}(i\omega) - \hat{G}_r(i\omega) \right\|_2}_{e_{in}}.$$

$$(3.33)$$

The spectral norm is invariant under orthogonal transformations. Consequently, the balanced truncation part (the error caused by the inner reduction e_{in}) follows from (3.19), (3.32), (3.33)

$$e_{in} = \left\| V_{O_{r_o}} G_r(s) V_{I_{r_i}}^T - V_{O_{r_o}} \tilde{G}_r(s) V_{I_{r_i}}^T \right\|_2 = \left\| G_r(s) - \tilde{G}_r(s) \right\|_2 \leq \delta.$$

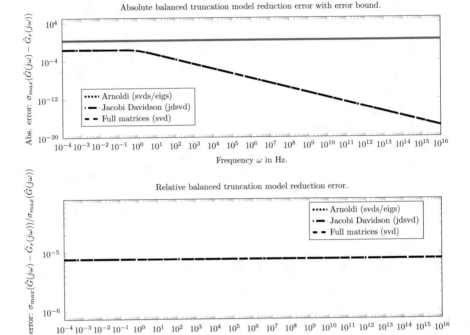

Fig. 3.11 Absolute and relative error of *eqs_lin*, $n = 14, 261$, $r = 300$ via 573 (Arnoldi and full SVD) and 520 (JDSVD) virtual terminals

To analyze the terminal reduction part, also called outer reduction error e_{out}, we start by considering RLC circuits only, i.e., $p = m$ denoted by m, $L = B^T$, and, if $s_0E - A \geq 0$, consequently $G(s) = G(s)^T$. Due to symmetry, $M_I = M_O = U\Sigma V^T$, and also $V_I = V_O = V$. Moreover $U = V$ holds in the SVDMOR case, which means that there is only one i-th s_0 frequency-shifted block moment in the ansatz matrices ($\kappa = 1$), e.g. $\mathbf{m}_0^0 \in \mathbb{R}^{m \times m}$, such that

$$M_I = M_O^T = \mathbf{m}_0^0 = B^T(s_0E - A)^{-1}B = U\Sigma V^T = U\Sigma U^T \approx U_r\Sigma_r U_r^T.$$

The local terminal reduction error e_{out} at s_0 then is

$$
\begin{aligned}
e_{out} = \left\| G(s_0) - \hat{G}(s_0) \right\|_2 &= \left\| B^T(s_0E - A)^{-1}B - U_rB_r^T(s_0E - A)^{-1}B_rV_r^T \right\|_2 \\
&\stackrel{(U=V)}{=} \left\| B^T(s_0E - A)^{-1}B - U_rU_r^TB^T(s_0E - A)^{-1}BU_rU_r^T \right\|_2 \\
&= \left\| U\Sigma U^T - U_rU_r^TU\Sigma U^TU_rU_r^T \right\|_2 = \left\| U\Sigma U^T - U_r\Sigma_r U_r^T \right\|_2 \\
&\stackrel{(SVD)}{=} \sigma_{r+1}^{I/O},
\end{aligned}
$$

if we keep r singular values or terminals. Applying balanced truncation, the total error at s_0 in the SVDMOR case in the spectral norm is then given by

$$e_{tot} \leq \sigma_{r+1}^{I/O} + 2 \sum_{j=n_r+1}^{n} \hat{\sigma}_j. \tag{3.34}$$

Hence, at least in the neighborhood of the expansion point chosen for the terminal reduction, we can expect an error of the same order of magnitude.

In the ESVDMOR case we allow $\kappa \geq 1$ (κ different $\mathbf{m}_i^0$ within the ansatz matrices). For simplicity let us assume $\kappa = 3$ ($\mathbf{m}_0^{s_0}$, $\mathbf{m}_1^{s_0}$, and $\mathbf{m}_2^{s_0}$) and $s_0 \in \mathbb{R}$. Thus,

$$M_I = \begin{pmatrix} \mathbf{m}_0^{s_0} \\ \mathbf{m}_1^{s_0} \\ \mathbf{m}_2^{s_0} \end{pmatrix} = \begin{pmatrix} U^{(1)} \\ U^{(2)} \\ U^{(3)} \end{pmatrix} \Sigma V = \begin{pmatrix} U_1^{(1)} & U_2^{(1)} \\ U_1^{(2)} & U_2^{(2)} \\ U_1^{(3)} & U_2^{(3)} \end{pmatrix} \begin{pmatrix} \Sigma_1 & 0 \\ 0 & \Sigma_2 \end{pmatrix} \begin{pmatrix} V_1^T \\ V_2^T \end{pmatrix}$$

$$=: \begin{pmatrix} U_1 & U_2 \end{pmatrix} \begin{pmatrix} \Sigma_1 & 0 \\ 0 & \Sigma_2 \end{pmatrix} \begin{pmatrix} V_1^T \\ V_2^T \end{pmatrix},$$

where the row partitioning in U is as in M_I, M_O and the column partitioning refers to the number of kept singular values, which we denote by the number r. We get $\mathbf{m}_j^0 = U^{(j)} \Sigma V^T, j = 1, 2, 3$, (which is not an SVD as $U^{(j)}$ is not orthogonal, but $\|U^{(j)}\|_2 \leq 1$ holds). Thus we can write

$$G(s) - \hat{G}(s) = \sum_{j=0}^{\infty} (\mathbf{m}_j^{s_0} - \hat{\mathbf{m}}_j^{s_0})(s - s_0)^j$$

$$= (\mathbf{m}_0^{s_0} - \hat{\mathbf{m}}_0^{s_0}) + (\mathbf{m}_1^{s_0} - \hat{\mathbf{m}}_1^{s_0})(s - s_0) + (\mathbf{m}_2^{s_0} - \hat{\mathbf{m}}_2^{s_0})(s - s_0)^2 + \mathcal{O}(s - s_0)^3.$$

Defining $P_1 := V_1 V_1^T$, and consequently $I - P_1 = V_2 V_2^T$, we can bound the first expressions thus as follows:

$$\mathbf{m}_j^{s_0} - \hat{\mathbf{m}}_j^{s_0} = \mathbf{m}_j^{s_0} - P_1 \mathbf{m}_j^{s_0} P_1 = U^{(j)} \Sigma V^T - P_1 U^{(j)} \Sigma V^T V_1 V_1^T$$

$$= U^{(j)} \begin{pmatrix} \Sigma_1 & 0 \\ 0 & \Sigma_2 \end{pmatrix} \begin{pmatrix} V_1^T \\ V_2^T \end{pmatrix} - P_1 U^{(j)} \begin{pmatrix} \Sigma_1 & 0 \\ 0 & \Sigma_2 \end{pmatrix} \begin{pmatrix} I_k \\ 0 \end{pmatrix} V_1^T$$

$$= U^{(j)} \begin{pmatrix} \Sigma_1 & 0 \\ 0 & 0 \end{pmatrix} \begin{pmatrix} V_1^T \\ V_2^T \end{pmatrix} +$$

$$+ U^{(j)} \begin{pmatrix} 0 & 0 \\ 0 & \Sigma_2 \end{pmatrix} \begin{pmatrix} V_1^T \\ V_2^T \end{pmatrix} - P_1 U^{(j)} \underbrace{\begin{pmatrix} \Sigma_1 \\ 0 \end{pmatrix} V_1^T}_{= \begin{pmatrix} \Sigma_1 & 0 \\ 0 & 0 \end{pmatrix} \begin{pmatrix} V_1^T \\ V_2^T \end{pmatrix}}$$

$$= U^{(j)} \begin{pmatrix} 0 & 0 \\ 0 & \Sigma_2 \end{pmatrix} V^T + (I - P_1) U^{(j)} \begin{pmatrix} \Sigma_1 \\ 0 \end{pmatrix} V_1^T$$

$$= U^{(j)} \begin{pmatrix} 0 & 0 \\ 0 & \Sigma_2 \end{pmatrix} V^T + V_2 V_2^T U^{(j)} \begin{pmatrix} \Sigma_1 \\ 0 \end{pmatrix} V_1^T =: e_{j,1} + e_{j,2}.$$

We can now express the error as follows:

$$G(s) - \hat{G}(s) = e_{0,1} + e_{1,1}(s - s_0) + e_{2,1}(s - s_0)^2$$
$$+ e_{0,2} + e_{1,2}(s - s_0) + e_{2,2}(s - s_0)^2 + \mathcal{O}(s - s_0)^3,$$

where, when taking norms, and using $\|U^{(j)}\|_2 \leq 1$, $\|V^T\|_2 = 1$,

$$\|e_{j,1}\|_2 \leq \sigma_{r+1}.$$

The terms $\|e_{j,2}\|_2$ cannot be bounded in a meaningful way. If σ_{r+1} is zero, then $V_2 V_2^T$ projects onto the nullspace of M_I, such that if σ_{r+1} is small enough, $V_2 V_2^T$ is still an orthoprojector onto the joint approximate nullspace of the first κ moments. That is, the error, up to order $\kappa - 1$, is essentially contained in the nullspace of the first κ moments. Future investigations may focus on exploiting this fact to get a general error bound.

3.4.3 Implementation Details

An efficient implementation is very important, especially if large scale systems are taken into account. Usually, within ESVDMOR we perform a full SVD and then set all singular values smaller than a threshold value equal to zero. This threshold value depends on the error we allow for ESVDMOR, see Sect. 3.4.2. The most obvious idea to increase the efficiency of the ESVDMOR approach is to avoid this full SVD in (3.14) because the ensuing step in the algorithm is to discard the expensively computed information. The numerical costs for an SVD of M_I can grow up to $\mathcal{O}(\kappa p m^2)$ flops. In case of M_O, these costs may be of order $\mathcal{O}(\kappa m p^2)$ flops. This is not appropriate for large-scale circuit systems. The computation of the reduced-order model only needs the leading block-columns $V_{I_{r_i}}$ and $V_{O_{r_o}}$ of the orthogonal matrices computed with the SVD. Consequently, we apply a truncated SVD, which can be computed cheaply employing sparsity of the involved matrices. There are a couple of truncated SVD approaches available, e.g., [12, 13, 46]. Efficient algorithms based on the Arnoldi method [31, 44] or the Lanczos method [4] as representatives of Krylov subspace methods are suggested in [6, 43]. A Jacobi-Davidson SVD approach is explained in [24]. Both possibilities will be explained in the following.

We start by having a look at joint features. Both methods are based on iterative computation of eigenvalues. Without loss of generality, we have a look at the input response moment matrix M_I. The same observations also hold for M_O. It is known that the computation of the eigenpairs of the $m \times m$ matrix $M_I^T M_I$ leads to the singular values of M_I [23]. But as forming the product should be avoided for reasons of numerical stability and efficiency (we even do not form M_I, see Sect. 3.4.3.3), both methods work with the augmented matrix

$$\mathscr{A}_I = \begin{bmatrix} 0 & M_I \\ M_I^T & 0 \end{bmatrix}. \tag{3.35}$$

It is shown, e.g. in [45], that the eigenvalues of $\mathscr{A}_I$ are of the form

$$\Lambda(\mathscr{A}_I) = \{-\sigma_m, \ldots, -\sigma_1, \sigma_1, \ldots, \sigma_m\}$$

and the positive eigenvalues are the singular values of M_I.

3.4.3.1 The Implicitly Restarted Arnoldi Method

One way to compute a TSVD is based on the implicitly restarted Arnoldi method [31, 44]. This approach exploits the symmetry of the matrix $\mathscr{A}_I$. Hence, it becomes equivalent to the symmetric Lanczos algorithm without taking advantage of the special block structure of $\mathscr{A}_I$. The Arnoldi method computes an orthonormal basis of a Krylov subspace and approximates eigenpairs for the projection of $\mathscr{A}_I$ onto the Krylov subspace. This iterative method becomes more expensive the more iterations it needs until convergence. This motivates the use of the restarted version, which is explained briefly in the following. A restart means that we delete parts of the calculated Krylov basis after a certain number of steps. Unfortunately, this also means a loss of information. In detail, the Arnoldi method after $j = k + l$ steps without restart gives

$$\mathscr{A}_I Q_j = Q_j H_j + q_{j+1} \left[0, \ldots, 0, h_{j+1,j} \right],$$
$$= Q_j H_j + h_{j+1,j} q_{j+1} e_j^T,$$

where Q_j is an orthogonal matrix and $H_j \in \mathbb{C}^{j \times j}$ is an upper Hessenberg matrix. The number of eigenvalues we are interested in is $k = 2r_i$ and l is the number of Krylov vectors we want to discard. Applying l QR steps with shifts $\theta_1, \ldots, \theta_l$, which are often chosen to be the l eigenvalues of H_j corresponding to the unwanted part of the spectrum, we get a matrix U such that

$$\hat{H}_j = U^* H_j U.$$

So we also have

$$\hat{Q}_j := Q_j U, \qquad u_j^T = e_j^T U.$$

With $f_{j+1} = h_{j+1,j} q_{j+1}$ we get

$$\mathscr{A}_l \hat{Q}_j = \hat{Q}_j \hat{H}_j + f_{j+1} u_j^T.$$

Now we split the components in

$$\hat{Q}_j = \begin{bmatrix} \hat{Q}_k & \tilde{Q}_l \end{bmatrix}, \qquad \hat{H}_j = \begin{bmatrix} \hat{H}_k & * \\ \beta e_1 e_k^T & \tilde{H}_k \end{bmatrix}, \qquad u_j^T = [0, \ldots, 0, \alpha, *, \ldots, *].$$

After removing the last l columns we end up with

$$\begin{aligned} \mathscr{A}_l \hat{Q}_k &= \hat{Q}_k \hat{H}_k + \beta \tilde{Q}_l e_1 e_k^T + f_{m+1}[0, \ldots, 0, \alpha], \\ &= \hat{Q}_k \hat{H}_k + (\beta \tilde{Q}_l e_1 + \alpha f_{m+1}) e_k^T, \\ &= \hat{Q}_k \hat{H}_k + \hat{f}_{m+1} e_k^T. \end{aligned}$$

We get again an Arnoldi recursion equivalent to the one we would get after k steps with starting vector $\hat{q}_1$. Until we reach an appropriate stopping criterion, see, e.g., [31], we perform l steps again and restart until the eigenvalues of $\hat{H}_k$ have converged to the largest singular values of M_l.

3.4.3.2 The Jacobi-Davidson SVD Method

A Jacobi-Davidson variant for singular value computation based on the augmented matrix $\mathscr{A}_l$ in (3.35) is proposed in [24]. The given block structure is exploited by the usage of two search spaces $\mathscr{U} \subset \mathbb{R}^{\kappa p}$, $\mathscr{V} \subset \mathbb{R}^m$ and respectively two test spaces $\mathscr{X} \subset \mathbb{R}^{\kappa p}$, $\mathscr{Y} \subset \mathbb{R}^m$, where κ is still the number of moment matrices of $G(s)$ of which the matrix M_l is constructed. We introduce matrices $U \in \mathbb{R}^{\kappa p \times k}$, $V \in \mathbb{R}^{m \times k}$, $X \in \mathbb{R}^{\kappa p \times k}$, $Y \in \mathbb{R}^{m \times k}$ whose columns are equal to the basis vectors of the four subspaces mentioned above.

For approximations to the singular values, that we call θ here, we use an auxiliary real number η and vectors $u \in \mathscr{U}$, $v \in \mathscr{V}$. Further, we employ the fact that for a singular triple (σ_i, u_i, v_i) of $M_l \in \mathbb{R}^{\kappa p \times m}$ the following equations

$$M_l v_i = \sigma_i u_i, \qquad M_l^T u_i = \sigma_i v_i \tag{3.36}$$

hold for $i = 1, \ldots, \text{rank}(M_l)$.

Definition 3.4.15 For vectors $x \in \mathbb{R}^m$, $y \in \mathbb{R}^n$ and subspaces $\mathscr{X} \subset \mathbb{R}^m$, $\mathscr{Y} \subset \mathbb{R}^n$, we say that the composed vector $\begin{pmatrix} x \\ y \end{pmatrix} \in \mathbb{R}^{m+n}$ is *double orthogonal* to $\begin{pmatrix} \mathscr{X} \\ \mathscr{Y} \end{pmatrix}$ if both $x \perp \mathscr{X}$ and $y \perp \mathscr{Y}$. We denote this by

$$\begin{pmatrix} x \\ y \end{pmatrix} \perp\!\!\!\perp \begin{pmatrix} \mathscr{X} \\ \mathscr{Y} \end{pmatrix}.$$

Similar to the orthogonal complement of a single vector we denote the subspace $\{u, v \in \mathbb{R}^m \times \mathbb{R}^n : x^T u = y^T v = 0\}$ as $(x, y)^{\perp\perp}$.

We impose a double Galerkin condition for the residual $res = res(\theta, \eta)$ defined as

$$res(\theta, \eta) := \begin{bmatrix} M_I v - \theta u \\ M_I^T u - \eta v \end{bmatrix} \perp\!\!\!\perp \begin{bmatrix} \mathscr{X} \\ \mathscr{Y} \end{bmatrix}. \tag{3.37}$$

Introducing $c, d \in \mathbb{R}^k$, such that $u = Uc$, $v = Vd$, (3.37) is equivalent to

$$\begin{cases} X^T M_I V d & = \theta X^T U c, \\ Y^T M_I^T U c & = \eta Y^T V d. \end{cases} \tag{3.38}$$

The assumption $x^T u \neq 0$, $y^T v \neq 0$ for test vectors $x \in \mathscr{X}$, $y \in \mathscr{Y}$ leads to approximations $\theta = \frac{x^T M_I v}{x^T u}$ and $\eta = \frac{y^T M_I^T u}{y^T v}$ for the singular values, that do not necessarily need to be equal. This depends on the choices for the test spaces $\mathscr{X}$, $\mathscr{Y}$, which is examined later. Suppose we already have approximations (θ, η, u, v) and look for new singular vector approximations $(\tilde{u}, \tilde{v}) = (u + s, v + t)$, which fulfill a double orthogonal correction $(s, t) \perp\!\!\!\perp (u, v)$, such that

$$M_I(v + t) = \sigma(u + s), \tag{3.39}$$
$$M_I^T(u + s) = \tau(v + t).$$

Equation (3.39) can be rearranged to

$$\begin{bmatrix} -\sigma I_{\kappa p} & M_I \\ M_I^T & -\tau I_m \end{bmatrix} \begin{bmatrix} s \\ t \end{bmatrix} = -res + \begin{bmatrix} (\sigma - \theta)u \\ (\tau - \eta)v \end{bmatrix}. \tag{3.40}$$

Since we are searching in $(x, y)^{\perp\perp}$ we consider the projection P onto $(x, y)^{\perp\perp}$ along (u, v), which expands (3.40) to

$$\underbrace{\begin{bmatrix} I_{\kappa p} - \frac{1}{x^T u} u x^T & 0 \\ 0 & I_m - \frac{1}{y^T v} v y^T \end{bmatrix}}_{\text{projection } P} \begin{bmatrix} -\theta I_{\kappa p} & M_I \\ M_I^T & -\eta I_m \end{bmatrix} \begin{bmatrix} s \\ t \end{bmatrix} = -res. \tag{3.41}$$

We also replace the unknown σ, τ by the known approximation θ, η in (3.41). Since $\forall \widetilde{x} \in \mathbb{R}^{\kappa p}, \forall \widetilde{y} \in \mathbb{R}^m$ with $u^T \widetilde{x} \neq 0$, $v^T \widetilde{y} \neq 0$,

$$
\begin{bmatrix} I_{\kappa p} - \frac{1}{u^T \widetilde{x}} \widetilde{x} u^T & 0 \\ 0 & I_m - \frac{1}{v^T \widetilde{y}} \widetilde{y} v^T \end{bmatrix} \begin{bmatrix} s \\ t \end{bmatrix} = \begin{bmatrix} s \\ t \end{bmatrix}
$$

holds, we can expand (3.41) to the JDSVD correction equation.

$$
\begin{bmatrix} I_{\kappa p} - \frac{1}{x^T u} u x^T & 0 \\ 0 & I_m - \frac{1}{y^T v} v y^T \end{bmatrix} \begin{bmatrix} -\theta I_{\kappa p} & M_I \\ M_I^T & -\eta I_m \end{bmatrix} \begin{bmatrix} I_{\kappa p} - \frac{1}{u^T \widetilde{x}} \widetilde{x} u^T & 0 \\ 0 & I_m - \frac{1}{v^T \widetilde{y}} \widetilde{y} v^T \end{bmatrix} \begin{bmatrix} s \\ t \end{bmatrix} = -res,
$$

$$(3.42)$$

with $(s, t) \perp\!\!\!\perp (u, v)$. If $\widetilde{x}$, $\widetilde{y}$ are nonzero multiples of x, y, the operator in (3.42) is symmetric and maps $(u, v)^{\perp\perp}$ to $(x, y)^{\perp\perp}$.

At this point, the Galerkin choice of the test spaces needs to be explained. If we choose $\mathscr{X} = \mathscr{U}$ and $\mathscr{Y} = \mathscr{V}$, with $\dim \mathscr{U} = \dim \mathscr{V} = k$, we get

$$
U^T M_I V d = \theta c, \qquad V^T M_I^T U c = \eta d
$$

for (3.38). Again, we have approximations $u = Uc$ and $v = Vd$. With test vectors $x = u, y = v$, and with $\|u\| = \|v\| = 1$, we have singular value approximations $\theta = \eta = u^T A v$ which are equal. Furthermore, for $\widetilde{x} = u$, $\widetilde{y} = v$ the correction equation (3.42) becomes

$$
\begin{bmatrix} I_{\kappa p} - u u^T & 0 \\ 0 & I_m - v v^T \end{bmatrix} \begin{bmatrix} -\theta I_{\kappa p} & M_I \\ M_I^T & -\theta I_m \end{bmatrix} \begin{bmatrix} I_{\kappa p} - u u^T & 0 \\ 0 & I_m - v v^T \end{bmatrix} \begin{bmatrix} s \\ t \end{bmatrix} = -res, \qquad (3.43)
$$

in which the operator is symmetric and maps $(u, v)^{\perp\perp}$ to itself. We conclude the standard variant of the JDSVD in Algorithm 3.1. The orthonormalization is performed by MGS and RMGS, which stands for the *modified Gram-Schmidt-* and the *repeated modified Gram-Schmidt-orthonormalization*. Due to orthogonality this Galerkin choice is optimal for computing of the largest singular values which is the problem we want to solve. Computing the singular values closest to a target τ is equivalent to computing the singular values of largest magnitude of the shifted and inverted matrix $(A - \tau I)^{-1}$, which implies a different correction equation.

A comparison of the influence of different decomposition methods applied within the terminal reduction of *RC549* to one terminal can be found in Fig. 3.12. The magnitude approximation is satisfying while the approximation of the phase becomes even worse by means of Arnoldi or JDSVD. If we keep 66 terminals,

---

Algorithm 3.1 JDSVD with standard Galerkin choice for computation of $\sigma_{max}(M_I)$

Input: Initial vectors (s, t), tolerance ϵ.
Output: Approximate singular triple (σ_{max}, u, v).
1: **for** $k = 1$ to ... **do**
2:     Orthonormalize s, t w.r.t. $\mathscr{U}_k$, $\mathscr{V}_k$.
3:     Expand search spaces $\mathscr{U}_k$, $\mathscr{V}_k$ with vectors s, t.
4:     Compute largest singular triple (θ, c, d) of $H_k = U_k^T M_I V_k$.
5:     Set $u = U_k c$, $v = V_k d$.
6:     Compute residual $res = \begin{bmatrix} M_I v - \theta u \\ M_I^T u - \theta v \end{bmatrix}$.
7:     **if** $\|res\| \leq \epsilon$ **then**
8:         return,
9:     **else**
10:        Approximately compute solution of correction equation (3.43) for $(s, t) \perp\!\!\!\perp (u, v)$.
11:    **end if**
12: **end for**

Bode plot of *RC549*, node 1,

original (blue, $n = 141$) and reduced system

(green, $r = 5$ via 1 virtual terminal).

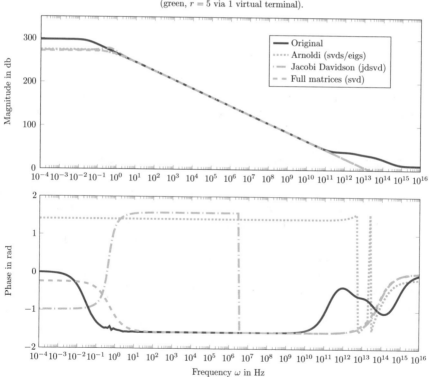

Fig. 3.12 Bode plot of node 1 of *RC549* reduced to 5 states via 1 virtual terminal

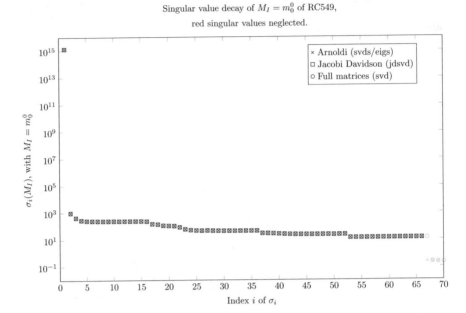

Fig. 3.13 Singular value decay of *RC549*, $M_I = m_0^0$, 66 virtual terminals

see Figs. 3.13 and 3.14, we can see a slight improvement. In particular, the phase approximation of the JDSVD approach becomes reasonable in the important frequency range. Like mentioned before, in this example the crucial point is the order reduction by means of BT, such that keeping more terminals does not significantly improve the overall approximation. Figure 3.15, as well as a zoomed version of it in Fig. 3.16, show that reducing the number of terminals of *eqs_lin* leads to a clear number of virtual terminals. Choosing machine precision as relative tolerance, the algorithms automatically truncate after 573 virtual terminals. The fact that JDSVD already stops after 520 virtual terminals is only a question of internal settings and does not express any disadvantage of the approach. If we need a very accurate terminal reduction, maybe it is worth the effort of calculating a whole SVD of the explicitly computed I/O-response moment matrices. If efficiency plays a role, both methods, Arnoldi and JDSVD, lead much faster to acceptable results.

3.4.3.3 Efficiency Issues

No matter which method we choose, the explicit computation of the response moment matrices, see Definitions 3.2.2 and 3.2.3, is too expensive and possibly numerically unstable. Moreover, using an established algorithm we need to provide

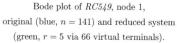

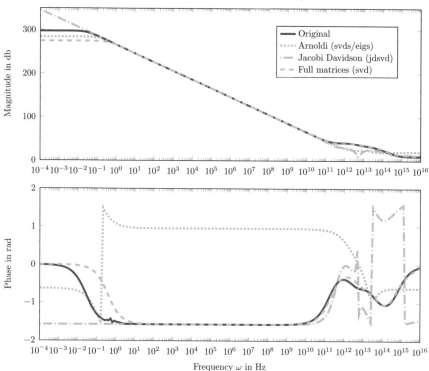

Fig. 3.14 Bode plot of node 1 of *RC549* reduced via 66 virtual terminals

the information of the matrix $\mathscr{A}_I$ in (3.35)

$$\mathscr{A}_I = \begin{bmatrix} 0 & M_I \\ M_I^T & 0 \end{bmatrix} \quad \text{or} \quad \mathscr{A}_O := \begin{bmatrix} 0 & M_O \\ M_O^T & 0 \end{bmatrix}$$

applied to a vector $\mathbf{x}$, i.e., $\mathscr{A}_{I/O}\mathbf{x} =: \mathbf{y}$, to build up the required subspaces. We use a function for this purpose. In the following we distinguish the two cases above.

3.4.3.4 Truncated SVD of the Input Response Moment Matrix M_I

Consider the augmented matrix $\mathscr{A}_I \in \mathbb{R}^{(\kappa p + m) \times (\kappa p + m)}$. The function input arguments are a vector $\mathbf{x} \in \mathbb{R}^{\kappa p + m}$ and a scalar κ, which is equal to the number of used moments κ, see (3.9). Output argument is a vector $\mathbf{y} \in \mathbb{R}^{\kappa p + m}$. We assume a block structure

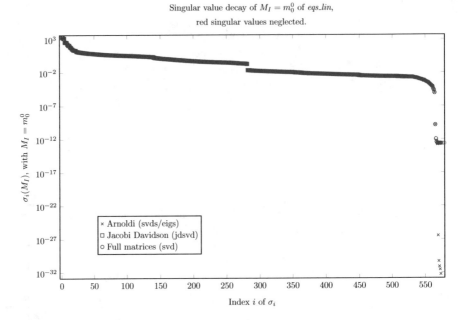

Fig. 3.15 Singular value decay of *eqs_lin*, $M_I = m_0^0$

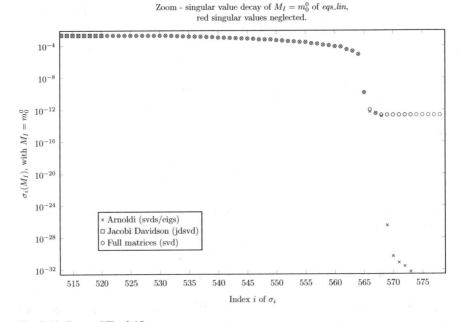

Fig. 3.16 Zoom of Fig. 3.15

of the vectors $\mathbf{x}$ and $\mathbf{y}$ corresponding to the block structure of M_I in $\mathscr{A}_I$, such that

$$
\begin{pmatrix} \mathbf{y}^0 \\ \mathbf{y}^1 \\ \vdots \\ \mathbf{y}^{\kappa-1} \\ \mathbf{y}^{\kappa} \end{pmatrix} = \begin{pmatrix} & & & \begin{bmatrix} \mathbf{m}_0 \\ \mathbf{m}_1 \\ \vdots \\ \mathbf{m}_{\kappa-1} \end{bmatrix} \\ & 0 & & \\ \begin{bmatrix} \mathbf{m}_0{}^T & \mathbf{m}_1{}^T & \cdots & \mathbf{m}_{\kappa-1}{}^T \end{bmatrix} & & 0 \end{pmatrix} \begin{pmatrix} \mathbf{x}^0 \\ \mathbf{x}^1 \\ \vdots \\ \mathbf{x}^{\kappa-1} \\ \mathbf{x}^{\kappa} \end{pmatrix}, \tag{3.44}
$$

where for $i = 0, \ldots, \kappa - 1$,

$$
\mathbf{y}^i = \begin{pmatrix} y_{ip+1} \\ \vdots \\ y_{(i+1)p} \end{pmatrix}, \quad \mathbf{y}^{\kappa} = \begin{pmatrix} y_{\kappa p+1} \\ \vdots \\ y_{\kappa p+m} \end{pmatrix}, \quad \mathbf{x}^i = \begin{pmatrix} x_{ip+1} \\ \vdots \\ x_{(i+1)p} \end{pmatrix}, \quad \text{and} \quad \mathbf{x}^{\kappa} = \begin{pmatrix} x_{\kappa p+1} \\ \vdots \\ x_{\kappa p+m} \end{pmatrix}.
$$

Performing the matrix-vector multiplication, we get the components $\mathbf{y}^i$ for $i = 0, \ldots, \kappa - 1$ as

$$
\mathbf{y}^i = \mathbf{m}_i \mathbf{x}^{\kappa} \quad \text{and} \quad \mathbf{y}^{\kappa} = \mathbf{m}_0{}^T \mathbf{x}^0 + \mathbf{m}_1{}^T \mathbf{x}^1 + \cdots + \mathbf{m}_{\kappa-1}{}^T \mathbf{x}^{\kappa-1} = \sum_i \mathbf{m}_i{}^T \mathbf{x}^i.
$$

To compute these components efficiently, we replace the block moments by their factors according to (3.2.2) and (3.2.3). Following Algorithm 3.2, we compute the parts of $\mathbf{y}$ by repeatedly applying the factors to parts of $\mathbf{x}$. We want to emphasize that we use the same factors each time. Providing this function to the methods described above enables the efficient computation of the desired singular values and vectors of M_I.

3.4.3.5 Truncated SVD of the Output Response Moment Matrix M_O

The computation of the truncated SVD of M_O works nearly analogously. Due to the fact that M_O has a different structure, also the structure in (3.44) changes to

$$
\begin{pmatrix} \mathbf{y}^0 \\ \mathbf{y}^1 \\ \vdots \\ \mathbf{y}^{\kappa-1} \\ \mathbf{y}^{\kappa} \end{pmatrix} = \begin{pmatrix} & & & \begin{bmatrix} \mathbf{m}_0{}^T \\ \mathbf{m}_1{}^T \\ \vdots \\ \mathbf{m}_{\kappa-1}{}^T \end{bmatrix} \\ & 0 & & \\ \begin{bmatrix} \mathbf{m}_0 & \mathbf{m}_1 & \cdots & \mathbf{m}_{\kappa-1} \end{bmatrix} & & 0 \end{pmatrix} \begin{pmatrix} \mathbf{x}^0 \\ \mathbf{x}^1 \\ \vdots \\ \mathbf{x}^{\kappa-1} \\ \mathbf{x}^{\kappa} \end{pmatrix}, \tag{3.45}
$$

Algorithm 3.2 Computation of components y^i, $i = 0, \ldots, \kappa$ (input response moment matrix)

Input: System matrices A, E, B, C, vector x, number of moment matrices κ, and frequency s_0.
Output: Vector $\mathbf{y} = \mathscr{A}_l \mathbf{x}$.
1: Initialize $y = 0$.
2: Compute the inverse $P = (s_0 E - A)^{-1}$, please see Remark 3.4.16.
3: % Prepare for loop
4: $a = B\mathbf{x}^\kappa$;
5: $a = Pa$;
6: **for** $i = 0$ to $(\kappa - 1)$ **do**
7:     % Computation of the first κ parts
8:     $\mathbf{y}^i = Ca$;
9:     % If necessary set up for the next iteration,
10:    **if** $i \neq (\kappa - 1)$ **then**
11:        $a = Ea$
12:        $a = Pa$
13:    **end if**
14:    % Compute factors, partly by embedded loop
15:    $b = C^T \mathbf{x}^i$;
16:    **for** $j = 0$ to $(i - 1)$ **do**
17:        $b = P^T b$;
18:        $b = E^T b$;
19:    **end for**
20:    $b = P^T b$;
21:    % Computation of the last part
22:    $\mathbf{y}^\kappa = \mathbf{y}^\kappa + B^T b$;
23: **end for**

where also the block structure of x and y changes, such that for $i = 0, \ldots, \kappa - 1$,

$$\mathbf{y}^i = \begin{pmatrix} y_{im+1} \\ \vdots \\ y_{(i+1)m} \end{pmatrix}, \quad \mathbf{y}^\kappa = \begin{pmatrix} y_{\kappa m+1} \\ \vdots \\ y_{\kappa m+p} \end{pmatrix}, \quad \mathbf{x}^i = \begin{pmatrix} x_{im+1} \\ \vdots \\ x_{(i+1)m} \end{pmatrix}, \quad \text{and} \quad \mathbf{x}^\kappa = \begin{pmatrix} x_{\kappa m+1} \\ \vdots \\ x_{\kappa m+p} \end{pmatrix}.$$

Applying matrix vector multiplication, we get again the components $\mathbf{y}^i$ for $i = 0, \ldots, \kappa - 1$ as follows, leading to Algorithm 3.3:

$$\mathbf{y}^i = \mathbf{m_i}^T \mathbf{x}^\kappa \quad \text{and} \quad \mathbf{y}^\kappa = \mathbf{m_0} \mathbf{x}^0 + \mathbf{m_1} \mathbf{x}^1 + \cdots + \mathbf{m_{\kappa-1}} \mathbf{x}^{\kappa-1} = \sum_i \mathbf{m_i} \mathbf{x}^i.$$

Remark 3.4.16 In line 2 of Algorithm 3.2 and of Algorithm 3.3, we require the computation of an inverse for ease of notation. In later lines it is easy to see that this inverse is applied in a matrix-vector product. Hence, the application of the inverse should be achieved by solving the corresponding linear system of equations, using, e.g., a pre-computed LU decomposition, or iterative methods.

Algorithm 3.3 Computation of components y^i, $i = 0, \ldots, \kappa$ (output response moment matrix)

Input: System matrices A, E, B, C, vector x, number of moment matrices κ, and frequency s_0.
Output: Vector $\mathbf{y} = \mathscr{A}_O \mathbf{x}$.
 1: Initialize $y = 0$.
 2: Compute the inverse $P = (s_0 E - A)^{-1}$, see Remark 3.4.16.
 3: % Prepare for loop
 4: $a = C^T \mathbf{x}^\kappa$;
 5: **for** $i = 0$ to $(\kappa - 1)$ **do**
 6:     % Computation of the first κ parts
 7:     $\mathbf{y}^i = P^T a$;
 8:     $\mathbf{y}^i = B^T \mathbf{y}^i$;
 9:     % If necessary set up for the next iteration,
10:     **if** $i \neq (\kappa - 1)$ **then**
11:         $a = P^T a$;
12:         $a = E^T a$;
13:     **end if**
14:     % Compute factors, partly by embedded loop
15:     $b = B \mathbf{x}^i$;
16:     $b = Pb$;
17:     **for** $j = 0$ to $(i - 1)$ **do**
18:         $b = Eb$;
19:         $b = Pb$;
20:     **end for**
21:     % Computation of the last part
22:     $\mathbf{y}^\kappa = \mathbf{y}^\kappa + Cb$;
23: **end for**

A few points remain to discuss. Looking at Sect. 3.4.3.2, we see that the information about $\mathbf{y} = M_I \mathbf{x}$ as well as $\mathbf{y} = M_O \mathbf{x}$ is sufficient within the JDSVD due to exploiting the block structure of $\mathscr{A}_I$. Hence, Algorithms 3.2 and 3.3 simplify accordingly.

For high numbers of κ both methods, Arnoldi as well as JDSVD, become numerically unstable. A reasonable number κ depends on the specific system and fortunately is often small in practice. Moreover, for linear circuits with the same number of inputs and outputs, mostly one moment of the transfer function, i.e., $\kappa = 1$, is sufficient.

The question how many virtual terminal $G_r(s)$ should have, i.e. how many singular values and vectors of M_I and M_O we need to compute, is partly answered in Sect. 3.4.2. This information influences the tolerances given to the algorithms for computing the truncated singular value decomposition explained above.

3.5 Summary and Outlook

Linear descriptor systems with more than just a handful of inputs and outputs appear in a lot of applications. This work motivates the need of MOR for this kind of systems within the context of IC design and circuit simulation. Three basic, widely different approaches, are presented, i.e., reduction based on SVDs revealing the I/O-behavior, a graph partitioning approach, and the interpolation-based construction of a reduced-order model from measurement data. Afterwards, a concrete method of the first type, the so-called ESVDMOR, is explained in detail. The preservation of important properties is shown as well as the approximation error is analyzed. Furthermore, special attention is given to an efficient truncated SVD implementation in the framework of ESVDMOR. The performance of the three proposed ESVDMOR implementations is illustrated using two test cases. In conclusion, one may say that in all three cases, the approximation of the magnitude of the frequency response can be achieved satisfactorily, while the phase approximation suffers from the additional truncation error in the truncated SVD approaches. Hence, if preservation of the phase is of importance, one has to invest more computational resources and use the full SVD of the generalized moment matrices.

Acknowledgements The authors want to thank Daniel Schmidthäusler, Sebastian Schöps and Georg Denk for their support. The work reported in this chapter was supported by the German Federal Ministry of Education and Research (BMBF), grant no. 03BEPAE1. Responsibility for the contents of this publication rests with the authors.

References

1. Anderson, B.D.O., Vongpanitlerd, S.: Network Analysis and Synthesis. Prentice Hall, Englewood Cliffs, NJ (1973)
2. Antoulas, A.C.: Approximation of Large-Scale Dynamical Systems. SIAM Publications, Philadelphia, PA (2005)
3. Antoulas, A.C., Ionita, A.C., Lefteriu, S.: On two-variable rational interpolation. Linear Algebra Appl. **436**(8), 2889–2915 (2012)
4. Baglama, J., Reichel, L.: Augmented implicitly restarted Lanczos bidiagonalization methods. SIAM J. Sci. Comput. **27**(1), 19–42 (2005). doi:http://dx.doi.org/10.1137/04060593X
5. Benner, P.: Advances in balancing-related model reduction for circuit simulation. In: Roos, J., Costa, L.R.J. (eds.) Scientific Computing in Electrical Engineering SCEE 2008. Mathematics in Industry, vol. 14, pp. 469–482. Springer, Berlin/Heidelberg (2010)
6. Benner, P., Schneider, A.: Model order and terminal reduction approaches via matrix decomposition and low rank approximation. In: Roos, J., Costa, L.R.J. (eds.) Scientific Computing in Electrical Engineering SCEE 2008. Mathematics in Industry, vol. 14, pp. 523–530. Springer, Berlin/Heidelberg (2010)
7. Benner, P., Schneider, A.: On stability, passivity, and reciprocity preservation of ESVDMOR. In: Benner, P., Hinze, M., ter Maten, J. (eds.) Model Reduction for Circuit Simulation. Lecture Notes in Electrical Engineering, vol. 74, pp. 277–288. Springer, Berlin/Heidelberg (2011)
8. Benner, P., Schneider, A.: Some remarks on a priori error estimation for ESVDMOR. In: Michielsen, B., Poirier, J.R. (eds.) Scientific Computing in Electrical Engineering SCEE 2010. Mathematics in Industry, vol. 16, pp. 15–24. Springer, Berlin/Heidelberg (2012)
9. Benner, P., Quintana-Ortí, E.S., Quintana-Ortí, G.: Parallel model reduction of large-scale linear descriptor systems via balanced truncation. In: Proceedings of the 6th International Meeting on High Performance Computing for Computational Science VECPAR'04, Valencia, 28–30 June 2004, pp. 65–78 (2004)

10. Benner, P., Mehrmann, V., Sorensen, D.C. (eds.): Dimension Reduction of Large-Scale Systems. Lecture Notes in Computational Science and Engineering, vol. 45. Springer, Berlin/Heidelberg (2005)
11. Benner, P., Hinze, M., ter Maten, E.J.W. (eds.): Model Reduction for Circuit Simulation. Lecture Notes in Electrical Engineering, vol. 74. Springer, Dordrecht (2011)
12. Berry, M.W., Pulatova, S.A., Stewart, G.W.: Algorithm 844: Computing sparse reduced-rank approximations to sparse matrices. ACM Trans. Math. Softw. **31**(2), 252–269 (2005). doi:http://doi.acm.org/10.1145/1067967.1067972
13. Chan, T.F., Hansen, P.C.: Computing truncated singular value decomposition least squares solutions by rank revealing QR-factorizations. SIAM J. Sci. Stat. Comput. **11**(3), 519–530 (1990)
14. Dai, L.: Singular Control Systems. Lecture Notes in Control and Information Sciences, vol. 118. Springer, Berlin (1989)
15. Feldmann, P., Freund, R.W.: Efficient linear circuit analysis by Padé approximation via the Lanczos process. In: Proceedings of EURO-DAC '94 with EURO-VHDL '94, Grenoble, pp. 170–175. IEEE Computer Society Press, Los Alamitos (1994)
16. Feldmann, P., Liu, F.: Sparse and efficient reduced order modeling of linear subcircuits with large number of terminals. In: ICCAD '04: Proceedings of the 2004 IEEE/ACM International Conference Computer-Aided Design, pp. 88–92. IEEE Computer Society, Washington, DC (2004)
17. Feldmann, P., Freund, R.W., Acar, E.: Power grid analysis using a flexible conjugate gradient algorithm with sparsification. Tech. rep., Department of Mathematics, UC Davis (2006). http://www.math.ucdavis.edu/~freund/flexible_cg.pdf
18. Francis, B.A.: A Course in H_∞ Control Theory. Lecture Notes in Control and Information Sciences. Springer, Berlin/Heidelberg (1987)
19. Freund, R.W.: Model reduction methods based on Krylov subspaces. Acta Numer. **12**, 267–319 (2003)
20. Freund, R.W.: SPRIM: structure-preserving reduced-order interconnect macromodeling. In: Proceedings of the 2004 IEEE/ACM International Conference on Computer-Aided Design, ICCAD '04, pp. 80–87. IEEE Computer Society, Washington, DC (2004)
21. Freund, R.W.: On Padé-type model order reduction of J-Hermitian linear dynamical systems. Linear Algebra Appl. **429**(10), 2451–2464 (2008)
22. Freund, R.W.: The SPRIM algorithm for structure-preserving order reduction of general RCL circuits. In: Benner, P., Hinze, M., ter Maten, J. (eds.) Model Reduction for Circuit Simulation. Lecture Notes in Electrical Engineering, vol. 74, pp. 25–52. Springer, Berlin/Heidelberg (2011)
23. Golub, G., Van Loan, C.: Matrix Computations, 3rd edn. Johns Hopkins University Press, Baltimore (1996)
24. Hochstenbach, M.E.: A Jacobi–Davidson type SVD method. SIAM J. Sci. Comput. **23**(2), 606–628 (2001)
25. Ionutiu, R.: Model order reduction for multi-terminal systems with applications to circuit simulation. Ph.D. thesis, Jacobs University, Bremen and Technische Universiteit Eindhoven, Eindhoven (2011)
26. Ionutiu, R., Rommes, J., Schilders, W.H.A.: SparseRC: Sparsity preserving model reduction for RC circuits with many terminals. IEEE Trans. Circuits Syst. **30**(12), 1828–1841 (2011)
27. Jain, J., Koh, C.K., Balakrishnan, V.: Fast simulation of VLSI interconnects. In: IEEE/ACM International Conference on Computer Aided Design, ICCAD-2004, pp. 93–98 (2004)
28. Kerns, K., Yang, A.: Stable and efficient reduction of large, multiport RC networks by pole analysis via congruence transformations. IEEE Trans. Circuits Syst. **16**(7), 734–744 (1997)
29. Lefteriu, S., Antoulas, A.C.: A new approach to modeling multiport systems from frequency-domain data. IEEE Trans. Comput. Aided Des. Integr. Circuits Syst. **29**(1), 14–27 (2010)
30. Lefteriu, S., Antoulas, A.C.: Topics in model order reduction with applications to circuit simulation. In: Benner, P., Hinze, M., ter Maten, E.J.W. (eds.) Model Reduction for Circuit Simulation. Lecture Notes in Electrical Engineering, vol. 74, pp. 85–107. Springer, Dordrecht (2011)

31. Lehoucq, R.B., Sorensen, D.C., Yang, C.: ARPACK Users' Guide: Solution of Large Scale Eigenvalue Problems with Implicitly Restarted Arnoldi Methods. Software, Environments, Tools, vol. 6, 142 p. SIAM, Society for Industrial and Applied Mathematics, Philadelphia, PA (1998)

32. Liu, P., Tan, S.X.D., Li, H., Qi, Z., Kong, J., McGaughy, B., He, L.: An efficient method for terminal reduction of interconnect circuits considering delay variations. In: ICCAD '05: Proceedings of the 2005 IEEE/ACM International Conference on Computer-Aided Design, pp. 821–826. IEEE Computer Society, Washington, DC (2005)

33. Liu, P., Tan, S.X.D., Yan, B., McGaughy, B.: An extended SVD-based terminal and model order reduction algorithm. In: Proceedings of the 2006 IEEE International Behavioral Modeling and Simulation Workshop, pp. 44–49 (2006)

34. Liu, P., Tan, S.X.D., McGaughy, B., Wu, L., He, L.: TermMerg: an efficient terminal-reduction method for interconnect circuits. IEEE Trans. Circuits Syst. 26(8), 1382–1392 (2007). doi:10.1109/TCAD.2007.893554

35. Liu, P., Tan, S.X.D., Yan, B., McGaughy, B.: An efficient terminal and model order reduction algorithm. Integr. VLSI J. 41(2), 210–218 (2008)

36. Lloyd, S.P.: Least squares quantization in PCM. IEEE Trans. Inform. Theory 28, 129–137 (1982)

37. Lozano, R., Maschke, B., Brogliato, B., Egeland, O.: Dissipative Systems Analysis and Control: Theory and Applications. Springer, New York, Inc., Secaucus, NJ (2000)

38. Mehrmann, V., Stykel, T.: Balanced truncation model reduction for large-scale systems in descriptor form. In: Benner, P., Mehrmann, V., Sorensen, D.C. (eds.) Dimension Reduction of Large-Scale Systems. Lecture Notes in Computational Science and Engineering, vol. 45, pp. 83–115. Springer, Berlin/Heidelberg (2005)

39. Odabasioglu, A., Celik, M., Pileggi, L.T.: PRIMA: passive reduced-order interconnect macro-modeling algorithm. IEEE Trans. Comput. Aided Des. Integr. Circuits Syst. 17(8), 645–654 (1998)

40. Reis, T.: Circuit synthesis of passive descriptor systems – a modified nodal approach. Int. J. Circuit Theory Appl. 38(1), 44–68 (2010)

41. Reis, T., Stykel, T.: Positive real and bounded real balancing for model reduction of descriptor systems. Int. J. Control 82(1), 74–88 (2010)

42. Schmidthausler, D., Schöps, S., Clemens, M.: Reduction of linear subdomains for non-linear electro-quasistatic field simulations. IEEE Trans. Magn. 49(5), 1669–1672 (2013). doi:10.1109/TMAG.2013.2238905

43. Schneider, A.: Matrix decomposition based approaches for model order reduction of linear systems with a large number of terminals. Diplomarbeit, Chemnitz University of Technology, Faculty of Mathematics (2008)

44. Sorensen, D.C.: Implicit application of polynomial filters in a k-step Arnoldi method. SIAM J. Matrix Anal. Appl. 13(1), 357–385 (1992)

45. Stewart, G.W., Sun, J.G.: Matrix Perturbation Theory. Academic Press, inc., Boston, MA (1990)

46. Stoll, M.: A Krylov-Schur approach to the truncated SVD. Linear Algebra Appl. 436(8), 2795–2806 (2012)

47. Tan, S.X.D., He, L.: Advanced Model Order Reduction Techniques in VLSI Design. Cambridge University Press, New York (2007)

48. Ye, Z., Vasilyev, D., Zhu, Z., Phillips, J.R.: Sparse implicit projection (SIP) for reduction of general many-terminal networks. In: Proceedings of the 2008 IEEE/ACM International Conference on Computer-Aided Design, ICCAD '08, pp. 736–743. IEEE Press, Piscataway, NJ (2008)

49. Zecevic, A.I., Siljak, D.D.: Balanced decompositions of sparse systems for multilevel parallel processing. IEEE Trans. Circuits Syst. I: Fundam. Theory Appl. 41(3), 220–233 (1994)

Chapter 4
Coupling of Numeric/Symbolic Reduction Methods for Generating Parametrized Models of Nanoelectronic Systems

Oliver Schmidt, Matthias Hauser, and Patrick Lang

Abstract This chapter presents new strategies for the analysis and model order reduction of systems of ever-growing size and complexity by exploiting the hierarchical structure of analog electronical circuits. Thereby, the entire circuit is considered as a system of interconnected subcircuits. Given a prescribed error-bound for the reduction process, a newly developed algorithm tries to achieve a maximal reduction degree for the overall system by choosing the reduction degrees of the subcircuits in a convenient way. The individual subsystem reductions with respect to their prescribed error-bound are then performed using different reduction techniques. Combining the reduced subsystems a reduced model of the overall system results. Finally, the usability of the new techniques is demonstrated on two circuit examples typically used in industrial applications.

4.1 Introduction

In order to avoid immense time and financial effort for the production of deficiently designed prototypes of *integrated circuits* (ICs), industrial circuit design uses mathematical models and simulations for predicting and analysing the physical behavior of electronical systems. Hence, redesigns and modifications of the systems can easily be carried out on a computer screen and tested by subsequent simulation runs. Thereby, analog circuits in general are modelled by systems of *differential-algebraic equations* (DAEs), which are composed of component characteristics and Kirchhoff laws.

The development in fabrication technology of ICs during the last years led to an unprecedented increase of functionality of systems on a single chip. Nowadays, ICs have hundreds of millions of semiconductor devices arranged in several layers

O. Schmidt • M. Hauser • P. Lang (✉)
Fraunhofer Institute for Industrial Mathematics ITWM (Fraunhofer Institut für Techno-und Wirtschaftsmathematik ITWM), Fraunhofer-Platz 1, 67663 Kaiserslautern, Germany
http://www.itwm.fraunhofer.de/abteilungen/systemanalyse-prognose-und-regelung/elektronik-mechanik-mechatronik/analog-insydes.html

© Springer International Publishing AG 2017
P. Benner (ed.), *System Reduction for Nanoscale IC Design*,
Mathematics in Industry 20, DOI 10.1007/978-3-319-07236-4_4

and low-level physical effects such as thermal interactions or electromagnetic radiation cannot be neglected anymore in order to guarantee a non-defective signal propagation. Mathematical models based on DAEs, however, have almost reached their limit and cannot model these effects accurately enough. Consequently, *distributed elements* for critical components such as semiconductor devices and transmission lines are used which yield supplementary model descriptions based on *partial differential equations* (PDEs), where also the spatial dependencies are taken into account. The coupling with DAEs modelling the remaining parts of the circuit then leads to systems of *partial differential-algebraic equations* (PDAEs). A spatial semidiscretization finally results in very high-dimensional systems of DAEs, thus rendering analysis and simulation tasks unacceptably expensive and time consuming.

Since design verification requires a large number of simulation runs with different input excitations, for the reasons mentioned above, *model order reduction* (MOR) becomes inevitable. Dedicated techniques in various areas of research have been developed among which the most popular ones are *numerical methods* taylored for linear systems. Besides these, there also exist *symbolic methods* [8, 10, 15, 19, 20], where *symbolic* means that besides the system's variables also its parameters are given as symbols instead of numerical values (see Sect. 4.1.1). They indeed are costly to compute, but allow deeper analytical insights into functional dependences of the system's linear and nonlinear behavior on its parameters by maintaining the dominant ones in their symbolic form. The basic idea behind these methods is a stepwise reduction of the original system by comparing its *reference solution* to the solution of the so far reduced system by using *error functions* which measure the difference between the two solutions. Since the resulting reduced system contains its parameters and variables in symbolic form, these methods can be seen as a kind of parametric model order reduction (pMOR). Compared to the standard parametric model order reduction techniques [4, 12], the symbolic ones can be additionally applied to nonlinear systems.

In order to avoid infeasibility of analysis and reduction of systems of ever-growing size and complexity, new strategies exploiting their hierarchical structure have been developed in the current research project. They further allow for a coupling of distinct reduction techniques for different parts of the entire circuits.

The corresponding algorithms have been implemented in *Analog Insydes* [1], the software tool for symbolic modeling and analysis of analog circuits, that is developed and distributed by the Fraunhofer ITWMin Kaiserslautern, Germany. It is based on the computer algebra system *Mathematica* [21].

The new approach has been successfully applied with significant savings in computation time to both a differential and an operational amplifier typically used in industry. The reduced models also proved to be very robust with regard to different inputs such as highly non-smooth pulse excitations. Thus, the aptitude of the new hierarchical model reduction algorithm to circuits of industrial size has been shown.

4.1.1 Symbolic Modeling of Analog Circuits

In the field of analog electronic circuits, there are different ways of modeling of the devices' behaviors. The approach *Analog Insydes* uses is the combination of Kirchhoff laws with symbolic device models to generate a symbolic system of differential-algebraic equations. As mentioned before, *symbolic* means here that besides the system's variables also its parameters are given as symbols instead of numerical values.

For a better understanding, consider the following circuit consisting of a voltage source V, a resistor R and a diode D.

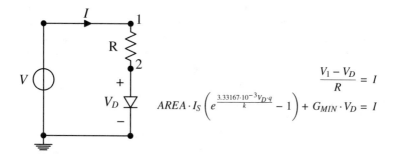

$$\frac{V_1 - V_D}{R} = I$$

$$AREA \cdot I_S \left(e^{\frac{3.33167 \cdot 10^{-3} V_D \cdot q}{k}} - 1 \right) + G_{MIN} \cdot V_D = I$$

The resulting system of equations contains the following equations modeling the current of the circuit by using the resistor's and diode's model equations. Additional to the system variables, like V_1, V_D and I, the parameters R, $AREA$, I_S, k, q and G_{MIN} are also given as symbols. This allows, besides the simulation after inserting the symbol's values, to analyse this system symbolically. That means in this case, that we could just solve symbolically the system for the voltage in node 1 with respect to the parameters and the voltage at the diode:

$$V_1 = R \cdot \left(AREA \cdot I_S \left(e^{\frac{3.33167 \cdot 10^{-3} V_D \cdot q}{k}} - 1 \right) + G_{MIN} \cdot V_D \right) + V_D$$

The next section follows the notes of [16–18].

4.2 Hierarchical Modelling and Model Reduction

In general, electronic circuits consist of a coupling of blocks such as amplifiers, current mirrors, or polarization circuits. Each block itself might have such a structure or is at least a network of interconnected components like diodes, resistors, transistors, etc. Consequently, the entire circuit is a hierarchical network of interconnected subcircuits, where each of these subcircuits may be modelled differently, e.g. based on netlists, PDEs, or DAEs.

The main idea behind the new *algorithm for hierarchical reduction* developed is the exploitation of the circuit's hierarchical structure in order to perform different reduction techniques on the distinct subcircuits. Besides a suitable choice of the methods according to the modelling of the corresponding subcircuits, this further allows for a faster processing of smaller subproblems if the administrative cost does not get out of hand. Furthermore, particularly in the case of symbolic model order reduction methods, like used in *Analog Insydes*, larger circuits become manageable at all.

Standard graph theoretical methods such as the *modified nodal analysis* (MNA) for transforming a circuit into a system of describing equations, however, lose the structural information available at circuit level. Therefore, we developed a new workflow for separate reductions of single subcircuits in the entire system, which uses information obtained from a previous simulation run. Since, in general, there is no relation between the errors of single nonlinear subsystems and the entire system available, we further introduced a new concept of *subsystem sensitivities*. By keeping track of the error on the output, which is resulting from the simplification of the subsystem, the sensitivities are used to measure the influence of single subsystems on the behavior of the entire circuit. Finally, these sensitivities are used to compute a *ranking of subsystem reductions*. In order to obtain a high degree of reduction for the entire system, it allows to replace the subcircuits by appropriate reduced models in an heuristically reasonable order. The details are explained in the following sections.

4.2.1 Workflow for Subsystem Reductions

Assume an electronic circuit Σ to be already hierarchically segmented into a set of m subcircuits T_i and an interconnecting structure S:

$$\Sigma = (\{T_i \mid i = 1, \ldots, m\}, S). \tag{4.1}$$

As already mentioned, each T_i itself might be recursively segmented into a set of subcircuits and a coupling structure. However, here we only consider a segmentation on the topmost "level 0". If one simply applies methods such as MNA to the circuit Σ in order to set up a set of describing equations, the resulting equations generally involve mixed terms from different subcircuits. In order to maintain the hierarchy information available on circuit level, in a first step the subcircuits are cut out from their connecting structure (cf. Fig. 4.1). Each subcircuit T is then connected to a test bench (a), i.e. a simulation test environment, where the voltage potentials at its terminals are recorded during a simulation run. For example, by simulating the original entire circuit, for each subcircuit T the interconnection of the remaining ones act as a test bench for T.

Note that the reduced model generated by the described method depends strongly on the input signals used. Thus, the input signal of the circuit has to cover the technical requirements of the later usage.

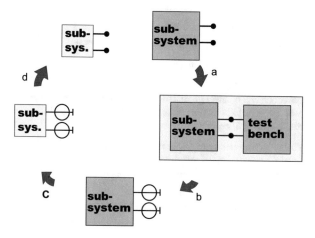

Fig. 4.1 Subsystem reduction via test bench approach

In a second step, the terminals of T are connected to voltage sources that generate exactly the recorded voltage potentials[1] (b). Hence, one has a closed circuit C_T with a defined input-output behavior at the terminals of T. A method such as MNA is used to set up a describing system F_T of equations[2] for C_T. Next, F_T can be reduced using arbitrary appropriate symbolic or numeric reduction techniques (c).

In a last step, the voltage sources at the terminals of the reduced model $\widetilde{F}_T$ are removed (d). Since the terminals of the subsystem are preserved during the reduction process, the original subcircuit T in Σ can easily be replaced by the reduced model $\widetilde{F}_T$ of F_T, thus using the same interconnecting structure S as introduced in (4.1). The entire procedure is repeated several times for each subcircuit T_i in Σ, thus yielding collections of reduced models for each T_i. The whole workflow is summarized in Algorithm 4.1.

It should further be mentioned here that this approach only controls the errors at the terminals of the single subcircuits. A priori, one cannot guarantee a certain *global error*, i.e. the error on the output of the entire circuit Σ, when replacing the original subcircuits T_i by reduced models $\widetilde{F}_{T_i}$. Thus the following algorithms were introduced to control the *global error* during the process.

---

[1] For doing it best, we first have to determine the voltage and current sources of the circuit that can act as inputs. Thus, the corresponding independent value of each port has to be considered as output. If you connect a voltage source at a port p this would be the current through port p, and vice versa.

For simplicity, we use here voltage sources as inputs and the currents as outputs. Besides of that, it turns out that residual based solvers simulate analog circuits containing transistors faster and more accurate if the voltages are given at the circuit's ports instead of the currents.

[2] Assume we are dealing with systems of DAEs. If PDEs are involved, apply a semidiscretization w.r.t. the spatial coordinates.

Algorithm 4.1 Reduction of subcircuits

Let $T = T_i$ be a subcircuit in an electronic circuit $\Sigma = (\{T_i \mid i = 1, \ldots, m\}, S)$.

a. Connect T to a test bench and record the voltage potentials at its terminals during a simulation run applying a suitable input.

b. Remove the test bench and connect grounded voltage sources to the terminals of T that generate exactly the recorded voltage potentials, thus having T isolated as a closed circuit C_T; further, set up a describing system of equations F_T for C_T.

c. Reduce F_T by using appropriate symbolic or numerical reduction techniques, where the voltages at all terminals of C_T are the inputs and the currents (flowing inwards) are the outputs. Here a family of reduced subsystems with different size and approximation quality is generated.

d. Remove the voltage sources at the terminals after the reduction and finally obtain a family of reduced subsystems, where each reduced subsystem $\widetilde{F}_T$ serves as a behavioral model of T.

4.2.2 Subsystem Sensitivities

In general, there is no relation between the error of the entire system and those of its nonlinear subsystems known. Therefore, in order to use reduced models of appropriate degree for the subsystems, in this section, we investigate the influence of single subcircuits T_i on the behavior of the entire circuit Σ given by (4.1). This offers a high degree of reduction also for Σ.

The goal here is to have an estimate of a subcircuit's *sensitivity*, i.e. the sensitivity of Σ with respect to changes in the corresponding subcircuit's behavior. Our novel approach measures the sensitivity by observing the influence of subcircuit reductions on the output of Σ and finally leads to a *ranking of subcircuit reductions*, i.e. an heuristically optimized order of subcircuit reductions.

Usually, the term *sensitivity analysis* in the background of electronic circuits means the influences of single components or system parameters on certain circuit or network variables. In that case, the *absolute sensitivity* of a variable z w.r.t. changes in a network parameter p is defined by

$$s_a(z, p) = \left. \frac{\partial z}{\partial p} \right|_{p = p_0}, \tag{4.2}$$

whereas

$$s_r(z, p) = p \left. \frac{\partial z}{\partial p} \right|_{p = p_0} = p \cdot s_a(z, p) \tag{4.3}$$

is the *relative sensitivity* of z w.r.t. p. In the two equations above, p_0 is the nominal value of p. Note that

$$s_a(z,p) \approx \left. \frac{\Delta z}{\Delta p} \right|_{p=p_0} = \frac{z - \widetilde{z}}{p_0 - \widetilde{p}} \tag{4.4}$$

is an approximation of s_a using perturbed values $\widetilde{z} = z(\widetilde{p})$ and $\widetilde{p}$ of $z = z(p)$ and $p = p_0$. While $z = z(p_0)$ corresponds to a simulation of Σ using the parameter $p = p_0$, $\widetilde{z}$ is obtained by using the perturbed parameter $p = \widetilde{p}$ during the simulation run.

Since we cannot derive the output y of Σ w.r.t. one of its subcircuits, we imitate the meaning of Eq. (4.4) by replacing a single subcircuit T in (4.1) by a perturbed version $\widetilde{T}$, i.e. by a reduced model $\widetilde{F}_T$ of its describing system of equations. Note that any other subsystem in Σ remains original, only T is replaced by one of its reduced models. We then simulate the configuration of Σ at hand and compare the *original* output y, i.e. the reference solution, to the perturbed entire system's output $\widetilde{y}$.

By Definition 4.2.1, the sensitivity of the subcircuit T in Σ is defined as the vector of tuples containing the reduced models and the resulting error on the perturbed entire system. For simplicity, we will not distinguish between sub*circuits* and the corresponding describing sub*systems* based on equations and denote both of them simply by T.

Definition 4.2.1 Let $\Sigma = (\{T_i \,|\, i = 1, \ldots, m\}, S)$ be an electronic circuit of interconnected subcircuits T_i connected by a structure S. Let further $T = T_i$ be one of the subcircuits in Σ. The **sensitivity of T in Σ** is the vector

$$s_T = \left((\widetilde{T}^{(1)}, E(y, y_{\widetilde{T}^{(1)}})), \ldots, (\widetilde{T}^{(m_T)}, E(y, y_{\widetilde{T}^{(m_T)}})) \right) \tag{4.5}$$

that contains tuples of **reduced models** $\widetilde{T}^{(j)}$ for T and the resulting error $E(y, y_{\widetilde{T}^{(j)}})$ on the original output y of Σ. In this notation, $y_{\widetilde{T}^{(j)}}$ is the output of the corresponding system

$$\Sigma_{\widetilde{T}^{(j)}} = \left(\{ \widetilde{T}^{(j)} \} \cup \{ T_i \,|\, i = 1, \ldots, m \} \setminus \{ T \}, S \right), \tag{4.6}$$

where T in comparison to the original circuit Σ is replaced by its jth reduced model $\widetilde{T}^{(j)}$.

In this definition, $\widetilde{T}^{(j)}$ denotes the jth reduced model of T which could be obtained by nonlinear symbolic model order reduction and an accepted error of 10% or by Arnoldi method and k iteration steps for example.

Note that the sensitivity of T involves systems $\Sigma_{\widetilde{T}^{(j)}}$ which are the same as Σ itself except for *exactly one subsystem*, namely T, that is replaced by a reduced version $\widetilde{T}^{(j)}$. Note further that these sensitivities depend again on the chosen input signals, as for the method introduced in Sect. 4.2.1.

Remarks 4.2.2 The sensitivity notion in Definition 4.2.1 can be further augmented by replacing the corresponding error $E(y, y_{\widetilde{T}^{(j)}})$ by a more general ranking expression that takes also additional subsystem criteria, like system size and sparsity, into account [9].

The next section describes how to use these sensitivities in order to obtain an heuristically reasonable order of subsystem reductions for the derivation of a system, that consists of reduced subsystems. Basically, the entries of the sensitivity vector of each subsystem are ordered increasingly with respect to the error on y. Then, following this order, the corresponding reduced models are used to replace the subsystems in Σ.

4.2.3 Subsystem Ranking

In this section, we present a strategy that allows an appropriate replacement of the subsystems of Σ by their reduced models in a reasonable order. The new algorithm presented here uses a ranking for deriving a hierarchically reduced model of the entire system Σ.

The basic idea behind the algorithm is ordering the reduced models of each subsystem increasingly w.r.t. the error[3] on the output y of Σ and subsequently performing the subsystem replacements according to this order. After each replacement, the accumulated error of the current subsystem configuration is checked by a simulation. If the user-given error bound ε for the error of the entire system Σ is exceeded, the current replacement is undone and the tested reduced model is deleted. Otherwise, the next replacement is performed and the procedure is repeated.

Let $\widetilde{T}_i^{(j)}$ denote the jth reduced model of the subsystem T_i. For each T_i in Σ we define a vector L_i which contains the entries of s_{T_i} and is increasingly ordered with respect to the error $E(y, y_{\widetilde{T}_i^{(j)}})$. The *original* subsystems T_i of Σ are then initialized by $\widetilde{T}_i^{(0)}$. In each iteration of the hierarchical reduction algorithm, the subsystem $\widetilde{T}_p^{(q)}$ that corresponds to the minimum entry[4] of the vectors L_i replaces the current (reduced) model $\widetilde{T}_p^{(q_0)}$ that is used for T_p in Σ. If the resulting accumulated error on the output y of Σ exceeds the user-specified error bound ε, the corresponding latest subsystem replacement is undone, i.e. $\widetilde{T}_p^{(q)}$ is reset to $\widetilde{T}_p^{(q_0)}$ in Σ. Furthermore, all reduced subsystems of subsystem T_p are deleted, since we assume that worse rated subsystems would also exceed the error bound. Otherwise only the corresponding sensitivity value $(\widetilde{T}_p^{(q)}, E(y, y_{\widetilde{T}_p^{(q)}}))$ of the tested reduced subsystem $\widetilde{T}_p^{(q)}$ is deleted from the vector L_p. This procedure is repeated until all the vectors L_i are empty. For a better overview of this approach see Algorithm 4.2.

---

[3]See Remarks 4.2.2.

[4]Minimal with respect to the corresponding error $E(y, y_{\widetilde{T}_i^{(j)}})$.

Algorithm 4.2 Heuristically reasonable order of subsystem replacements

Input: segmented electronic circuit $\Sigma = \big(\{T_i \mid i = 1, \ldots, m\}, S\big)$, input u, error bound ε

Output: reduced entire system $\widetilde{\Sigma} = \big(\{\widetilde{T}_i^{(j^*)} \mid i = 1, \ldots, m\}, S\big)$, where $\widetilde{T}_i^{(j^*)}$ are suitably reduced subsystems, $E(y, y_{\widetilde{\Sigma}}) \leq \varepsilon$, and where $y_{\widetilde{\Sigma}}$ is the output of $\widetilde{\Sigma}$

1: **for all** subsystems T_i **do**
2:     $L_i := \text{order}(s_{T_i})$ w.r.t. $E(y, y_{\widetilde{T}_i^{(j)}})$
3:     $\widetilde{T}_i^{(0)} := T_i$
4: **end for**

5: $\underset{\sim}{L} := (L_1, \ldots, L_m)$ $\triangleright$ set starting point
6: $\widetilde{\Sigma} := \Sigma$

7: $y := \text{solve}(\Sigma, u)$ $\triangleright$ calculate reference

8: **while** $L = \emptyset$ **do**
9:     compute $(\widetilde{T}_p^{(q)}, E(y, y_{\widetilde{T}_p^{(q)}})) := \min\limits_{i, L_i \in L}(\min(L_i))$ w.r.t. $E(y, y_{\widetilde{T}_i^{(j)}})$ $\triangleright$ choose reduced subsystem
10:     replace current $\widetilde{T}_p^{(q_0)}$ by $\widetilde{T}_p^{(q)}$
11:     update$(\widetilde{\Sigma})$ $\triangleright$ update and solve new reduced overall system
12:     $y_{\widetilde{\Sigma}} := \text{solve}(\widetilde{\Sigma}, u)$
13:     $\varepsilon_{\text{out}} := E(y, y_{\widetilde{\Sigma}})$
14:     delete[5] entry $(\widetilde{T}_p^{(q)}, E(y, y_{\widetilde{T}_p^{(q)}}))$ in L_p

15:     **if** $\varepsilon_{\text{out}} \leq \varepsilon$ **then** $\triangleright$ check resulting error
16:         **if** $dimension(L_p) = 0$ **then**
17:             delete[5] entry L_p in L
18:         **end if**
19:     **else**
20:         reset $\widetilde{T}_p^{(q)}$ to $\widetilde{T}_p^{(q_0)}$ $\triangleright$ undo reduction if error exceeds error bound
21:         update$(\widetilde{\Sigma})$
22:         delete[5] entry L_p in L
23:     **end if**

24: **end while**

Remarks 4.2.3 Note that Algorithm 4.2 can further be improved, e.g. by a clustering of subsystem replacements, where reduced models that cause a similar error on y are bundled in a cluster. Thus, costly multiple simulations for computing the solution $\widetilde{y}$ of the so far reduced entire system $\widetilde{\Sigma}$ are avoided, since they are performed only once after a whole cluster of subsystem replacements is executed. In case the error bound is still not violated, we can continue with the next cluster of subsystem

---

[5] For a vector $X = (x_1, \ldots, x_n)$, deleting the entry x_i in X means, that a vector $\widetilde{X} = (x_1, \ldots, x_{i-1}, x_{i+1}, \ldots, x_n)$ of dimension $n - 1$ results.

replacements. Otherwise, however, all replacements in the current cluster have to be rejected and it has to be subdivided for further processing.

Another idea for further improvements is the use of approximate simulations such as *k-step solvers* which quit the Newton iteration for computing the system's solution after k steps. Thus, one obtains an approximate solution $\widehat{y} \approx \widetilde{y}$ for the output of the so far reduced system $\widetilde{\Sigma}$ which can be used for the error check $E(y, \widehat{y}) \leq \varepsilon$ instead of $\widetilde{y}$.

4.2.4 Algorithm for Hierarchical Model Reduction

To combine all the considerations of the preceding sections, the *algorithm for hierarchical model reduction* exploiting the hierarchical structure of electronic circuits is set up. It is schematically shown in Fig. 4.2.

Remarks 4.2.4 Since electronic circuits even nowadays are designed in a modular way using building blocks of network devices and substructures such as current mirrors and amplifying stages, the hierarchical segmentation of an electronic circuit is given in a more or less natural way. Otherwise, the segmentation has to be made manually or by using *pattern matching* approaches[13] in order to detect substructures in the entire circuit.

Note that the presented algorithm (cf. Fig. 4.2) can be applied recursively to the subcircuit levels such that a hierarchically model order reduction results.

4.3 Implementations

The algorithms of the preceding sections have been completely implemented in *Analog Insydes* [1] and the approach for hierarchical model reduction was fully automated. It is divided into three main procedures

- ReduceSubcircuits,
- SensitivityAnalysis, and
- HierarchicalReduction

that have to be executed sequentially. Each of the above procedures takes several arguments among which there are some optional ones.

ReduceSubcircuits is called with the specification of an already segmented netlist of the circuit which is to be hierarchically reduced, the specification of the reduction method for each subcircuit, the simulation time interval necessary for recording the voltage potentials at the ports of the subcircuits, and several optional parameters. In accordance with the provided data, the procedure then computes the reduced models for all the specified subcircuits and appends them to the original

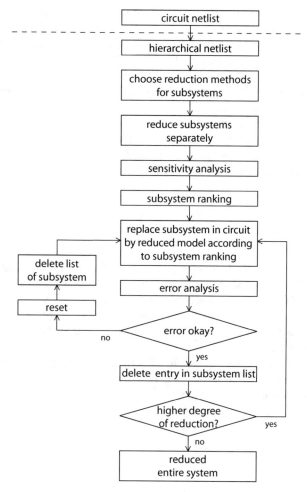

Fig. 4.2 Schematic illustration of the full algorithm for hierarchical model reduction using subsystem sensitivities.

circuit object. This offers an easy switching among the respective models for a single subcircuit.

The return value of ReduceSubcircuits, i.e. the hierarchically segmented circuit object together with the reduced models of each subcircuit, is then used as parameter of the function SensitivityAnalysis. In addition, the names of the reduced models, a specification of the output variables, the simulation time interval for the error check, and the error function itself to measure the error on the reference solution y are provided. The procedure computes the sensitivity vectors of each subcircuit and returns them ordered increasingly w.r.t. the error on y.

Finally, HierarchicalReduction needs a specification of the entire circuit and its reduced subcircuit models, the global error bound, the output variables, the

sensitivities returned by `SensitivityAnalysis`, the simulation time interval
necessary for the error check, and several optional arguments. Then the subsystem
replacements are performed according to the sensitivities and the accumulated error
is checked after each replacement (Algorithm 4.2). The procedure terminates when
all sensitivity lists have been processed and deleted.

In addition to the above, there have been implemented several data structures and
operators for their manipulation, as well as some well-known reduction algorithms,
transmission line models—based on a discretization of a PDE model—and further
components based on general state space systems. We further implemented some
environments to test the above procedures and functionalities. However, we will not
go into detail here, for an overview we refer to [16].

4.4 Applications

In order to demonstrate the large potential of the new hierarchical reduction
approach, it is applied in time domain to two analog circuit examples that are typical
representatives of components used in industrial circuit design. The results of the
hierarchical reduction of the two circuits are compared to the direct non-hierarchical
approach. Furthermore, some additional input excitations are applied to the circuits
in order to show the robustness of the derived reduced models.

Note that we present here the application of the introduced methods on circuits
containing strongly nonlinear devices to demonstrate the ability of the approach in
the field of nonlinear analog circuits.

4.4.1 Differential Amplifier

The differential-amplifier circuit shown in Fig. 4.3 consists of five subcircuits
DUT, DUT 2, L 1, L 8, and L 9, where the latter three ones are transmission lines
connecting the supply voltage sources VCC and VEE and the input voltage source
V1 with the remaining parts of the circuit. For the modelling of the transmission
lines, we take a discretized PDE model, namely, the telegrapher's equations (cf.,
e.g., [5–7, 11]), with 20 line segments each. While VCC and VEE generate constant
voltage potentials of $12\,V$ and $-12\,V$, respectively, the input voltage generated by
V1 is a sine wave excitation with an amplitude of $2\,V$ and a frequency of $100\,kHz$.
Finally, the computations are performed on a time interval $\mathbb{I} = [0.\,s, 10^{-5}\,s]$.

Using MNA to set up a system of describing DAEs yields 167 equations
containing 645 terms (on "level 0"). A *non-hierarchical* symbolic reduction of the
entire system then needs approximately 2 h and 11 min,[6] where most of that time

---

[6]The computations are performed on a Dual Quad Xeon E5420 with 2.5 MHz and 16 GB RAM.

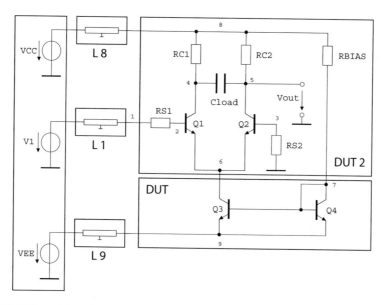

Fig. 4.3 Differential amplifier with its intuitive hierarchical segmentation into five subcircuits DUT, DUT 2, L 1, L 8, and L 9.

($\approx$95%) is needed for the computation of the transient *term ranking*.[7] Due to this, the computational costs are approximately the same for all choices of the error bound ε. The error function used first discretizes the time interval $\mathbb{I}$ to a uniform grid of 100 points and then takes the maximum absolute difference of the two solutions on this grid as a measure for the error.

With ε equal to 3% the system is reduced to 124 equations and 416 terms, while a permitted error of 10% narrows these numbers down to 44 equations and 284 terms. The results are shown in Fig. 4.4. Note also that the error bound of 10% is fully exploited.

In contrast to the immense time costs of the non-hierarchical approach, the new algorithm for hierarchical reduction reduces the entire system in only 4 min and 50 s. The subcircuits DUT and DUT 2 are reduced symbolically by using a sweep of error bounds

$$sw = \{1\%, 10\%, 50\%, 90\%, 100\%\}, \tag{4.7}$$

such that each subsystem yields 5 reduced subsystems. The three transmission lines L 1, L 8, and L 9 are reduced numerically by applying Arnoldi's algorithm [2, 3].

---

[7]A term ranking is a trade-off between accuracy and efficiency in computation time that estimates the influence of a term in a system of equations on its solution. Here, however, we use full simulations instead of low-accuracy estimates. For more details see [20].

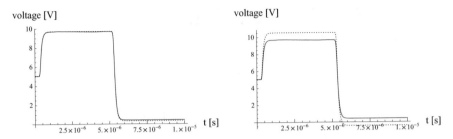

Fig. 4.4 Solution of the original (*solid*) and the non-hierarchically reduced system (*dotted*) allowing 3% (*left*) and 10% (*right*) maximum error, respectively. The input V1 is $2 \cdot Sin(2\pi 10^5 t)$ Volts

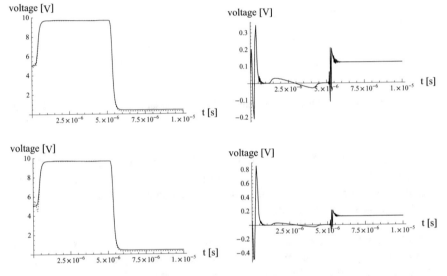

Fig. 4.5 *Left:* Solution of the original (*solid*) and the reduced system (*dotted*) allowing 3% (first row) and 10% (second row) maximum error, respectively. *Right:* The corresponding error plots. The input V1 is $2 \cdot Sin(2\pi 10^5 t)$ Volts

For L 1 there are five reduced models computed by performing the Arnoldi iteration for up to 5 steps, and for L 8 and L 9 there are made only up to 3 steps, thus yielding three reduced models each for L 8, and L 9.

For $\varepsilon = 3\%$ the resulting reduced overall system contains 62 equations with 315 terms, and $\varepsilon = 10\%$ leads to a reduced overall system with 60 equations and 249 terms. The solutions of the original and the respective reduced systems are shown in Fig. 4.5 together with the corresponding error plots.

In this case we conclude that the hierarchical reduction approach is more than 26 times faster than the non-hierarchical one. Also the number of equations of the reduced model in the 3% error case could be halved. Moreover, by applying further input excitations to both the original and the hierarchically reduced system with

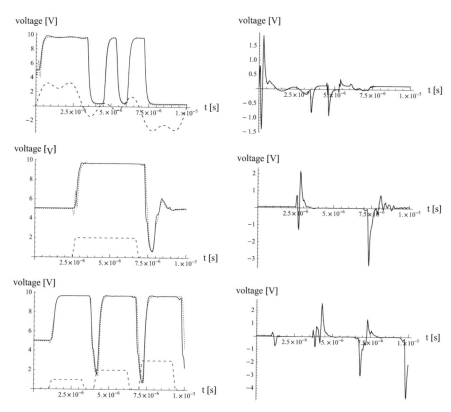

Fig. 4.6 *Left:* Solution of the original (*solid*) and the reduced system (*dotted*, $\varepsilon = 3\%$) together with the input excitation (*dashed*). *Right:* The corresponding error plots

$\varepsilon = 3\%$, it turns out that the derived model is very robust, even w.r.t. highly non-smooth pulse excitations (cf. Fig. 4.6). Note further that the simulation is accelerated approximately by a factor of 5.

4.4.2 Reduction of the Transmission Line L 1 by Using an Adapted PABTEC Algorithm

The tool *PABTEC* [14] uses the Balanced Truncation reduction technique to reduce the linear parts of an analog circuit. Please refer to Chap. 2.6 for further informations about this software.

To demonstrate the coupling of the introduced algorithm with a numeric model order reduction method, we use *PABTEC* to reduce the linear transmission line L 1. The remaining subcircuits DUT, DUT 2, L 8, and L 9 have been reduced by the same methods shown in the example before. In doing so, the original entire system

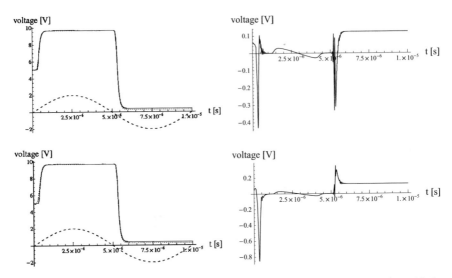

Fig. 4.7 *Left:* Solution of the original (*solid*) and the reduced system (*dotted*) together with the input excitation (*dashed*). *Right:* The corresponding error plots. The first row corresponds to the reduced system obtained by allowing an error of $\varepsilon = 3\%$, while the second row shows the results for $\varepsilon = 10\%$. The input V1 is $2 \cdot Sin(2\pi 10^5 t)$ Volts

consists of 191 equations containing 695 terms. Applying the hierarchical reduction algorithm with error bounds $\varepsilon = 3\%$ and $\varepsilon = 10\%$ then needs about 8 min and 20 s and yields systems with 96 equations and 2114 terms and 84 equations and 1190 terms, respectively. The results of their simulation (speed-up by a factor of approximately 5) are shown in Fig. 4.7.

4.4.3 Operational Amplifier

The second circuit example to which we apply the new algorithms is the operational amplifier op741 shown in Fig. 4.8. It contains 26 bipolar junction transistors (BJT) besides several linear components and is hierarchically segmented into seven subcircuits CM1–3, DP, DAR, LS, and PP. For a detailed description of their functionality in the interconnecting structure we refer to [16, Appendix C].

The goal is a *symbolic* reduction of the entire circuit in time domain with an overall error bound of $\varepsilon = 10\%$. While the input voltage source Vid provides a sine wave excitation of 0.8 V and 1 kHz frequency on a time interval $\mathbb{I} = [0\,\text{s}, 0.002\,\text{s}]$ to the system, its output is specified by the voltage potential of node 26. The input together with the corresponding output, i.e. the *reference solution*, is shown in Fig. 4.9. Note that the reference solution is pulse-shaped and, thus, the *standard* error function used for the differential amplifier in the preceding sections may lead to large errors for small delays in jumps of the solution. Hence, even with a

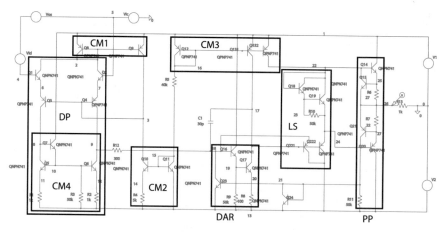

Fig. 4.8 Operational amplifier op741 composed of seven subcircuits CM1–3, DP, DAR, LS, PP

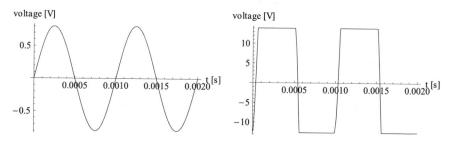

Fig. 4.9 Input voltage excitation (*left*) and the corresponding reference solution (*right*) of the operational amplifier op741

prescribed error bound of 10%, the system might not be reduced at all. In order to cope with these problems, here we use the $\mathscr{L}^2$-norm as error function.

Using MNA to set up a system of describing DAEs for the entire system yields 215 equations and 1050 terms. The direct non-hierarchical symbolic reduction method needs more than 10.5 h and yields a system containing 97 equations and 593 terms. At the same time, providing a sweep of error bounds

$$sw = \{2\%,\ 10\%,\ 20\%,\ 30\%,\ 50\%,\ 70\%,\ 90\%,\ 100\%\} \qquad (4.8)$$

for the separate symbolic reduction of all seven subcircuits and applying the hierarchical reduction algorithm needs only 2 h and 22 min. The resulting system, however, consists of 153 equations and 464 terms, which can be narrowed down to

Fig. 4.10 Output of the original (*solid*) and the hybrid reduced entire system (*dotted*)

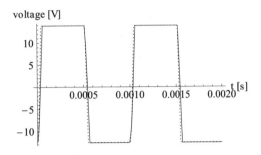

Table 4.1 Overview of the results of the reduction of the operational amplifier op741

Original system:		215 equations, 1050 terms, 26.0 s simulation time		
Error fct.		Non-hierarchical	Hierarchical	Hybrid
$\mathscr{L}^2$-norm	Time costs	10.5 h	2.5 h	<4 h
	Equations/terms	97/593	139/362	34/92
	Error	2.51%	7.16%	5.68%
	Simulation time	16.0 s	11.4 s	2.2 s
E^*	Time costs	>12 h	2.5 h	<4 h
	Equations/terms	80/405	132/336	34/93
	Error	0.37%	0.08%	5.32%
	Simulation time	9.5 s	13.1 s	2.0 s

The computations were performed on a machine with 8 Quad-core AMD Opteron 8384 "Shanghai" (32 cores in total) with 2.7 GHz and 512 GB RAM on a SuSE Linux 10.1 system

139 equations and 362 terms by slight manual improvements[8] of the hierarchical reduction algorithm.

Considering the obtained systems as *interim solutions* and applying a second non-hierarchical symbolic reduction then reduces the size drastically and leads to a model with only 34 equations and 92 terms. Simultaneously, there are almost no further changes for the non-hierarchically reduced system with 97 equations. Note that the additional time cost is less then 1.5 h, while the simulation time of the "*hybrid*" reduced model is significantly decreased.

Figure 4.10 offers a qualitative impression of the results obtained by the *hybrid* approach. Furthermore, earlier results involved a newly designed alternative error function E^* which is less sensitive with respect to small delays in jumps of the system's solution.

Table 4.1 provides an overview of the best results obtained by the three different approaches. See also Fig. 4.11 which offers some details about the accuracy, time

---

[8]Due to the structure preserving reduction method, the resulting reduced model contains equations connecting the models of the subcircuits, that can be avoided, like: Voltage of node 24 of subcircuit LS is equal to the voltage of node 24 of subcircuit PP.

Unifying the corresponding variables (i.e. $V\$24\$LS$ and $V\$24\$PP$) yields a decrease of the number of equations.

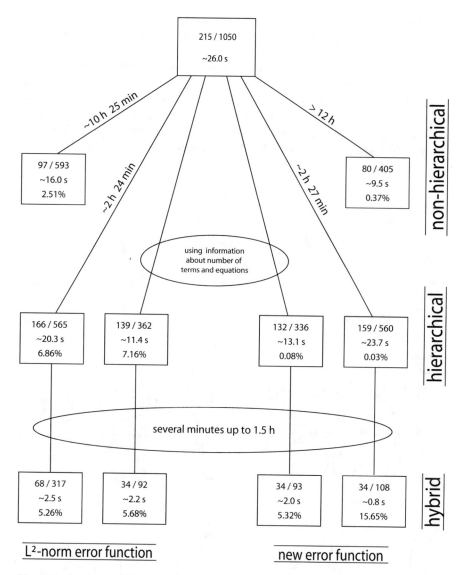

Fig. 4.11 Summary of the reduced models of the op741 amplifier obtained by the three different reduction approaches. The *boxes* contain the number of equations/terms of the reduced models, the time costs of a simulation using the original sine wave excitation, and the error on the output V$26 of the original amplifier

costs for simulation, and number of equations and terms of the different reduced models. We will not go into detail here, for further information we refer to [16] instead.

With a view towards the robustness of the derived models, we apply some further input excitations, namely, a sine wave with 3 kHz frequency, a sum of sine waves of 250, 500, and 2000 Hz, and a pulse excitation of 250 Hz. In addition to almost perfectly coinciding output curves of the corresponding reduced models (cf. Fig. 4.12), the speed-up in simulation time is up to a factor of 19, see Table 4.2. The presented systems are identified by their number of equations and terms.

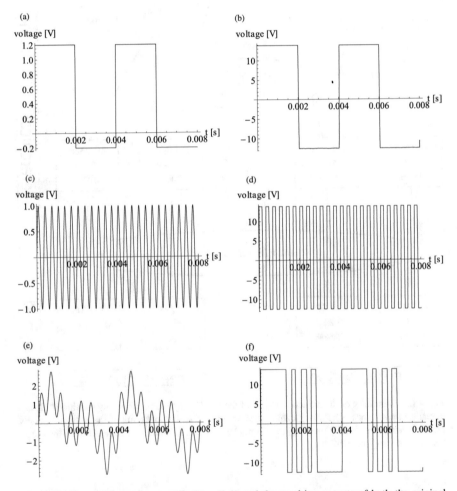

Fig. 4.12 Three different input excitations (*left*) and the resulting outputs of both the original (*solid*) and the *hybrid* reduced system (*dashed*). (**a**) A voltage pulse. (**b**) Output results for the voltage pulse. (**c**) A sine wave with frequency 3000 Hz. (**d**) Outputs applying the input in (**c**). (**e**) A sum of sine waves. (**f**) The outputs for the sum of sine waves

Table 4.2 Speed-up of simulation of a hybrid reduced entire system w.r.t. the original one

System	Voltage pulse	3 kHz Sine wave	Sum of sine waves
215 = 1050	106 s	273 s	104 s
34 = 92	6 : 6 s	14 : 1 s	10 : 5 s

4.5 Conclusions

To conclude this chapter, we briefly summarize the results: The new hierarchical reduction approach offers enormous savings in computation time, a significant speed-up in system simulations, and yields good reduced models w.r.t. the error, the number of equations and terms of the original system. Moreover, even for highly non-smooth pulse excitations, the reduced models turn out to be very robust. The developed methods were applied to two model classes, circuits consisting of nonlinear subcircuits and circuits containing subcircuits modelled by PDEs, that demonstrated the large potential of the new algorithms.

References

1. Analog Insydes website: http://www.itwm.fraunhofer.de/abteilungen/systemanalyse-prognose-und-regelung/elektronik-mechanik-mechatronik/analog-insydes.html
2. Antoulas, A.C.: Approximation of Large-Scale Dynamical Systems. SIAM, Philadelphia, PA (2005)
3. Bai, Z.: Krylov subspace techniques for reduced-order modeling of large-scale dynamical systems. Appl. Numer. Math. **43**, 9–44 (2002)
4. Feng, L., Benner, P.: A robust algorithm for parametric model order reduction. Proc. Appl. Math. Mech. **7**(1), 1021501–1021502 (2007)
5. Grabinski, H.: Theorie und Simulation von Leitbahnen. Signalverhalten auf Leitungssystemen in der Mikroelektronik. Springer, Berlin (1991)
6. Günther, M.: A Joint DAE/PDE model for interconnected electrical networks. Math. Comput. Model. Dyn. Syst. **6**(2), 114–128 (2000)
7. Günther, M.: Partielle differential-algebraische Systeme in der numerischen Zeitbereichs-analyse elektrischer Schaltungen. Fortschritt-Berichte VDI, Reihe 20, Nr. 343. VDI-Verlag, Düsseldorf (2001)
8. Halfmann, T., Wichmann, T.: Symbolic methods in industrial analog circuit design. In: Anile, A., Alì, G., Mascali, G. (eds.) Scientific Computing in Electrical Engineering. Mathematics in Industry, vol. 9, pp. 87–92. Springer, Heidelberg (2006)
9. Hauser, M., Salzig, C.: Hierarchical model-order reduction for robust design of parameter-varying systems. In: Symbolic and Numerical Methods, Modeling and Applications to Circuit Design (SMACD) (2012)
10. Hennig, E.: Symbolic Approximation and Modeling Techniques for Analysis and Design of Analog Circuits. Ph.D. Thesis, Universität Kaiserslautern, Shaker, Aachen (2000)
11. Miri, A.M.: Ausgleichsvorgänge in Elektroenergiesystemen. Springer, Berlin (2000)
12. Panzer, H., Mohring, J., Eid, R., Lohmann, B.: Parametric model order reduction by matrix interpolation. Automatisierungstechnik **58**, 475–484 (2010)

13. Pelz, G., Roettcher, U.: Pattern matching and refinement hybrid approach to circuit comparison. IEEE Trans. Comput.-Aided Des. Integr. Circuits Syst. **13**, 264–276 (1994)
14. Reis, T., Stykel, T.: PABTEC: passivity-preserving balanced truncation for electrical circuits. IEEE Trans. Comput.-Aided Des. Integr. Circuits Syst. **29**, 1354–1367 (2010)
15. Sommer, R., Hennig, E., Nitsche, G., Schwarz, P., Broz, J.: Automatic Nonlinear Behavioral Model Generation Using Symbolic Circuit Analysis. In: Fakhfakh, M., Tlelo-Cuautle, E., Fernández, F.V. (eds.) Design of Analog Circuits through Symbolic Analysis, pp. 305–341. Bentham Science Publishers, Sharjah (2012). doi:10.2174/978160805095611010303
16. Schmidt, O.: Structure-Exploiting Coupled Symbolic-Numerical Model Reduction for Electrical Networks. Dissertation, Technische Universität Kaiserslautern, Cuvillier, Göttingen (2010)
17. Schmidt, O.: Structure-exploiting symbolic-numerical model reduction of nonlinear electrical circuits. In: Proceedings der 16th European Conference on Mathematics for Industry ECMI 2010, Wuppertal. Mathematics in Industry, vol. 17, pp. 179–185. Springer, Berlin (2010)
18. Schmidt, O., Halfmann, T., Lang, P.: Coupling of numerical and symbolic techniques for model order reduction in circuit design. In: Benner, P., Hinze, M., ter Maten, E.J.W. (eds.) Model Reduction for Circuit Simulation. Lecture Notes in Electrical Engineering, vol. 74, pp. 261–275. Springer, Dordrecht (2011)
19. Sommer, R., Krauße, D., Schäfer, E., Hennig, E.: Application of symbolic circuit analysis for failure detection and optimization of industrial integrated circuits. In: Fakhfakh, M., Tlelo-Cuautle, E., Fernández, F.V. (eds.) Design of Analog Circuits through Symbolic Analysis, pp. 445–477. Bentham Science Publishers, Sharjah (2012). doi:10.2174/978160805095611201010305
20. Wichmann, T.: Symbolische Reduktionsverfahren für nichtlineare DAE-Systeme. Shaker, Aachen (2004)
21. Wolfram, S.: The Mathematica Book, 5th edn. Wolfram Media, Inc., Champaign (2003)

Chapter 5
Low-Rank Cholesky Factor Krylov Subspace Methods for Generalized Projected Lyapunov Equations

Matthias Bollhöfer and André K. Eppler

Abstract Large-scale descriptor systems arising from circuit simulation often require model reduction techniques. Among many methods, Balanced Truncation is a popular method for constructing a reduced order model. In the heart of Balanced Truncation methods, a sequence of projected generalized Lyapunov equations has to be solved. In this article we present a general framework for the numerical solution of projected generalized Lyapunov equations using preconditioned Krylov subspace methods based on iterates with a low-rank Cholesky factor representation. This approach can be viewed as alternative to the LRCF-ADI method, a well established method for solving Lyapunov equations. We will show that many well-known Krylov subspace methods such as (F)GMRES, QMR, BICGSTAB and CG can be easily modified to reveal the underlying low-rank structures.

5.1 Introduction

The numerical simulation of large-scale integrated circuits nowadays approaches system sizes of several hundred million equations. This ever-increasing size has several sources; one of which is the accelerating scale of miniaturization, another reason is the increasing density of the integrated devices. The simulation of the complete system requires many simulation runs with different input signals. These simulation runs would be impossible to compute in acceptable time using the original system. Instead it is necessary to replace the original system by a significantly smaller reduced model which inherits the essential structures and properties of the original system as, e.g., passivity and stability. To deal with this problem model order reduction techniques (MOR) have turned out to be a key technology in order to generate reduced models. Among the most popular methods

---

M. Bollhöfer (✉) • A.K. Eppler
TU Braunschweig, Carl-Friedrich-Gauß-Fakultät, Institut Computational Mathematics,
Universitätspl. 2, 38106 Braunschweig, Germany
e-mail: m.bollhoefer@tu-bs.de; a.eppler@tu-bs.de

© Springer International Publishing AG 2017
P. Benner (ed.), *System Reduction for Nanoscale IC Design*,
Mathematics in Industry 20, DOI 10.1007/978-3-319-07236-4_5

for MOR are those based on Krylov subspace method or *Balanced Truncation (BT)* [3, 11, 30]. For problems arising from circuit simulation in particular passivity-preserving balanced truncation methods [33, 34, 45] are of particular interest, since beside reducing the circuit to a reduced order model, major important properties like stability and passivity have to be preserved to obtain a physically correct model (see also Chaps. 2 and 3). Another frequently used method mainly applied to *partial differential-algebraic equations (PDAE)* is the *Proper Orthogonal Decomposition (POD)* method, cf. [16, 26], Chap. 1.

This article is organized as follows. In Sect. 5.2 we will give a brief introduction to balanced truncation which is the motivation for our methods and requires to solve several sequences of generalized projected Lyapunov equations. This includes existing numerical methods for solving Lyapunov equations. In Sect. 5.3 we will present our novel approach for generalized projected Lyapunov equations based on Krylov subspace methods. Finally we will use several examples from circuit simulation, as well as other examples, to demonstrate our approach in Sect. 5.4.

5.2   Balanced Truncation

The basis for the numerical methods for generalized projected Lyapunov equations presented in this paper are those using Balanced Truncation (BT). In particular passivity-preserving Balanced Truncation methods will be of special interest for model order reduction techniques applied to circuit simulation problems.

5.2.1   *Introduction to Balanced Truncation*

To start with the idea of Balanced Truncation we consider a linear time invariant descriptor system

$$\begin{aligned} E\dot{x} &= Ax + Bu \\ y &= Cx + Du \end{aligned} \quad \text{where } A, E \in \mathbb{R}^{n,n}, B \in \mathbb{R}^{n,m}, C \in \mathbb{R}^{p,n}, D \in \mathbb{R}^{p,m}$$

such that $m, p \ll n$. Numerical methods for MOR replace E, A, B, C by smaller matrices $\tilde{E}, \tilde{A}, \tilde{B}, \tilde{C}$ such that for all matrices the initial dimension n is replaced by a suitable $l \ll n$, i.e., $\tilde{A}, \tilde{E} \in \mathbb{R}^{l,l}, \tilde{B} \in \mathbb{R}^{l,m}, \tilde{C} \in \mathbb{R}^{p,l}$.

When using Balanced Truncation the reduction of the model is done by multiplying with matrices $W \in \mathbb{R}^{l,n}$, $T \in \mathbb{R}^{n,l}$ in order to obtain the reduced descriptor system

$$(E, A, B, C, D) \rightarrow (\tilde{E}, \tilde{A}, \tilde{B}, \tilde{C}, D) = (WET, WAT, WB, CT, D).$$

The transformation matrices W and T are constructed using the solutions of generalized Lyapunov equations, the so-called proper controllability gramian G_{pc} and the proper observability gramian G_{po}. When E is singular one also has to take into account the improper controllability gramian and the improper observability gramian, for details we refer to [34]. For the computation of a reduced model we have to compute $X = G_{pc}$ and $Y = G_{po}$ by solving the projected generalized Lyapunov equations

$$EXA^T + AXE^T + P_l BB^T P_l^T = 0, \text{ where } X = P_r X P_r^T,$$
$$E^T YA + A^T YE + P_r^T C^T C P_r = 0, \text{ where } Y = P_l^T Y P_l. \tag{5.1}$$

Here P_l, P_r are obtained from the Weierstrass canonical form for (E, A). To do so we assume that $\det(A - \lambda E) \neq 0$. In this case there exist nonsingular V and Z such that

$$V^{-1} EZ = \begin{pmatrix} I & 0 \\ 0 & N \end{pmatrix}, \ V^{-1} AZ = \begin{pmatrix} J & 0 \\ 0 & I \end{pmatrix}. \tag{5.2}$$

Here J, N denote matrices in Jordan canonical form where N nilpotent. The left and right projection of (E, A) to

$$(P_l EP_r, P_l AP_r) = \left(V \begin{pmatrix} I & 0 \\ 0 & 0 \end{pmatrix} Z^{-1}, \ V \begin{pmatrix} J & 0 \\ 0 & 0 \end{pmatrix} Z^{-1} \right) \tag{5.3}$$

yields the projectors P_l and P_r of the matrix pencil $\lambda E - A$ with respect to the subspace of finite eigenvalues. By solving (5.1) we obtain symmetric, positive semidefinite solutions $X = RR^T$ and $Y = LL^T$, provided that the eigenvalues of J from (5.2) are located in the open left half plane. In many application problems for MOR in circuit simulation the matrices X, Y are numerically of approximate low rank. Using the singular value decomposition of $L^T ER$ and $L^T AR$ the balanced system is built. This way some general properties such as passivity are not necessarily preserved. To even preserve passivity it is necessary to solve the projected Lur'e equations [33], see also Chap. 2. In some special cases these in turn can be traced back to algebraic Riccati equations of the form

$$EXA^T + AXE^T + (EXC^T - P_l B)^T R^{-1} (EXC^T - P_l B) = 0, \text{ where } X = P_r X P_r^T \tag{5.4}$$

and

$$A^T YE + E^T YA + (B^T YE - CP_r)^T R^{-1} (B^T YE - CP_r) = 0, \text{ where } Y = P_l^T Y P_l. \tag{5.5}$$

For details we refer to [33, 34]. Solving Riccati equations using Newton's method or the Newton-Kleinman method [4, 44] (cf. also Sect. 2.5.2) requires solving a sequence of projected, generalized Lyapunov equations of the form

$$
\begin{aligned}
EX_kA_k^T + A_kX_kE^T + P_lB_kB_k^TP_l^T &= 0, \text{ where } X_k = P_rX_kP_r^T, \\
E^TY_kA_k + A_k^TY_kE + P_r^TC_k^TC_kP_r &= 0, \text{ where } Y_k = P_l^TY_kP_l.
\end{aligned}
\tag{5.6}
$$

Compared with the original pencil (E, A), the matrix A_k in (E, A_k) is obtained from a low-rank correction of A. For large-scale sparse systems arising from circuit simulation this allows for the computation of sparse approximations (resp. sparse factorizations) of (E, A) and then to transfer these approximations to the pencil (E, A_k) using the Sherman–Morrison–Woodbury formula [14] with respect to A_k.

5.2.2 Numerical Methods for Projected, Generalized Lyapunov Equations

We will now describe in detail how projected, generalized Lyapunov equations of type

$$
EXA^T + AXE^T + P_lBB^TP_l^T = 0, \text{ where } X = P_rXP_r^T
\tag{5.7}
$$

are solved numerically. For simplicity we restrict ourselves to solving a single equation of this type which is at the heart of Balanced Truncation methods and in practice such equations have to be solved frequently, e.g. once per iteration in Algorithm 2.7.

One of the most commonly used methods for solving (projected) generalized Lyapunov equations is the ADI method [28, 31, 44, 47]. The ADI method for solving (5.7) consists of a sequence $j = 1, 2, 3, \ldots$ of steps, which is decomposed into two half-steps

$$
\begin{aligned}
(E + \tau_jA)X_{j-\frac{1}{2}}A^T &= -P_lBB^TP_l^T - AX_{j-1}(E - \tau_jA)^T, \\
AX_j(E + \tau_jA)^T &= -P_lBB^TP_l^T - (E - \tau_jA)X_{j-\frac{1}{2}}A^T.
\end{aligned}
$$

From these two coupled equations we successively compute $(X_j)_j$. Here $\tau_1, \tau_2, \tau_3, \ldots$ refer to shift parameters that have to be chosen appropriately to achieve convergence, see [32, 47]. Starting with $X_0 = 0$ and using that the right hand side $P_lBB^TP_l^T$ is symmetric and positive semidefinite one can easily verify that all iterates $X_j = R_jR_j^T$ are also symmetric and positive semidefinite. This can be used explicitly in the ADI method to represent the iterates by low rank Cholesky

---

Algorithm 5.1 LRCF-ADI for generalized, projected Lyapunov equations (5.7)

1: Compute shift parameters $\tau_1, \ldots, \tau_t$
2: $z_1 = \sqrt{-2\mathrm{Re}(\tau_1)}(E + \tau_1 A)^{-1} P_l B$
3: $R = [z_1]$
4: **for** $i = 2 \ldots t \ldots$ **do**
5:     $z_i = P_{i-1} = \frac{\sqrt{-2\tau_i}}{\sqrt{-2\tau_{i-1}}}\left[z_{i-1} - (\tau_i + \bar{\tau}_{i-1})(E + \tau_i A)^{-1} A z_{i-1}\right]$
6:     $R_i = [R_{i-i}\, z_i]$
7: **end for**

---

factors

$$R_j = \left[\sqrt{-2\mathrm{Re}(\tau_j)}\{(E + \tau_j A)^{-1} P_l B\},\ \{(E + \tau_j A)^{-1}(E - \bar{\tau}_j A) R_{j-1}\}\right].$$

For the generalized case, the projectors P_l and P_r from (5.3) ensure that if $R_{j-1} = P_r R_{j-1}$, then we also obtain $R_j = P_r R_j$ and thus $X_j = P_r X_j P_r^T$ holds.

The matrices of type $(E \pm \tau_j A)$, $(E \pm \tau_j A)^{-1}$ commute with each other independent on the choice of τ_j. This observation has been used in [28] to reduce the numerical complexity of the computation of R_j by one order of magnitude. This has lead to the *Low-Rank Cholesky Factor-ADI Method (LRCF-ADI)* and can be described for the case of general and projected Lyapunov equations by Algorithm 5.1.

For the convergence of the ADI method the choice of the shift parameters $\tau_1, \tau_2, \ldots$ is essential. For the case where $E = I$ and $-A$ is symmetric and positive definite optimal shift parameters are known [47]. In general one often has to work with heuristic parameters as, e.g., in [31, 32] although asymptotically optimal shifts can be determined by Fejér-Walsh points [43] or Leja-Bagby points [27, 42]. Also, recent global optimization strategies to approximate optimal shifts have lead to promising results [38].

5.3 Low-Rank Cholesky Factor Krylov Subspace Methods

The objective of this article is to describe novel numerical solution methods for projected generalized Lyapunov equations based on low-rank Krylov subspace methods. These are frequently used as core part of the model order reduction approach. In principle ADI methods belong to the class of iterative methods for solving the linear system (5.7). This can be equivalently rewritten as

$$\mathscr{L}\mathscr{X} = \mathscr{B},$$

where $\mathscr{L} = E \otimes A + A \otimes E$ corresponds to the Lyapunov operator in (5.1), $\mathscr{X} = \text{vec}(X)$ and, $\mathscr{B} = \text{vec}(-P_l BB^T P_l^T)$. Our goal is to preserve the matrix structure as well as the low-rank structure of the Lyapunov equation (5.7), while at the same time the benefits of structure-preserving preconditioned Krylov subspace methods applied to $\mathscr{L}\mathscr{X} = \mathscr{B}$ will be exploited.

5.3.1 Low-Rank Krylov Subspace Methods

Krylov subspace methods without preconditioning consist of series of matrix-vector multiplications, scalar products and linear combinations of vectors. The residuals $\mathscr{R}_k = \mathscr{B} - \mathscr{L}\mathscr{X}_k$ are located in $\text{span}\{\mathscr{R}_0, \mathscr{L}\mathscr{R}_0, \ldots, \mathscr{L}^{k-1}\mathscr{R}_0\}$ and the approximate solutions $\mathscr{X}_{k+1}$, respectively, can be represented by elements of the space $\mathscr{X}_0 + \text{span}\{\mathscr{R}_0, \mathscr{L}\mathscr{R}_0, \ldots, \mathscr{L}^{k-1}\mathscr{R}_0\}$. For two-sided Krylov subspace methods such as BiCG or QMR, multiplications with the transposed matrix also have be taken into account. Here as part of the solution process, both Riccati equations (5.4), (5.5) could be treated simultaneously solving both associated linear equations (5.6) in common. This follows from the property of two-sided Lanczos methods which require a right initial guess such as $P_l B_k B_k^T P_l^T$ and an appropriate left initial guess which could be chosen as $P_r^T C_k^T C_k P_r$. Yet the two-sided methods have to be slightly modified to explicitly compute the additional approximate solution. While the iterates are located in a Krylov subspace on one hand, on the other hand we have that the right hand side $-P_l BB^T P_l^T$ of the Lyapunov equation, as well as the approximate solution $X = RR^T$, can be represented as symmetric low-rank matrices. The obvious approach to migrate both structures for adapted Krylov subspace methods consists of keeping all iterates of the Krylov subspace method in symmetric low-rank format. This in turn yields elementary operations for iterates of type $Z_i = Q_i M_i Q_i^T$, where $M_i = M_i^T$, $i = 1, 2$ are also symmetric but of much smaller size than Z_i. We set $\mathscr{Z}_i = \text{vec}(Z_i)$ and note that elementary operations are translated as follows:

- $\mathscr{L}\mathscr{Z}_1$ is equivalently written as

$$EZ_1 A^T + AZ_1 E^T = \underbrace{[EQ_1, AQ_1]}_{=:Q_2} \underbrace{\begin{bmatrix} 0 & M_1 \\ M_1 & 0 \end{bmatrix}}_{=:M_2} \underbrace{[EQ_1, AQ_1]^T}_{=:Q_2^T}$$

- analogously, $\mathscr{L}^T \mathscr{Z}_1$ is represented by

$$E^T Z_1 A + A^T Z_1 E = \underbrace{[E^T Q_1, A^T Q_1]}_{=:Q_2} \underbrace{\begin{bmatrix} 0 & M_1 \\ M_1 & 0 \end{bmatrix}}_{=:M_2} \underbrace{[E^T Q_1, A^T Q_1]^T}_{=:Q_2^T}$$

- linear combinations $\alpha \mathscr{Z}_1 + \beta \mathscr{Z}_2$ can be traced back to

$$\alpha Z_1 + \beta Z_2 = \underbrace{[Q_1, Q_2]}_{=:Q_3} \underbrace{\begin{bmatrix} \alpha M_1 & 0 \\ 0 & \beta M_2 \end{bmatrix}}_{=:M_3} \underbrace{[Q_1, Q_2]^T}_{=:Q_3^T}$$

- finally, scalar products are easily computed using the trace of matrices by

$$\mathscr{Z}_1^T \mathscr{Z}_2 = \text{trace}(Z_1^T Z_2) = \text{trace}(Z_1 Z_2).$$

This shows that in principle Krylov subspace methods can be set up such that all iterates are represented by symmetric low-rank matrices.

5.3.2 Low-Rank Cholesky Factor Preconditioning

If we wish to supplement a Krylov subspace solver with an additional preconditioner, then in the worst case the low-rank structure of the single iterates is lost. This holds even for the simple example of diagonal preconditioning. Instead the preconditioner has to be adapted such that the low-rank structure is inherited. The natural choice for a preconditioner in this case is obtained from the LRCF-ADI method. Given $Z_1 = Q_1 M_1 Q_1^T$, we can apply t steps of the LRCF-ADI method from Sect. 5.2.2 starting with a right hand side Cholesky factor $B := Q_1$. This way we obtain the LRCF-ADI factors $(R_j)_{j=1,\dots,t}$ which in turn yield a symmetric low-rank matrix

$$\begin{array}{c} \text{LRCF-ADI} \\ Q_1 M_1 Q_1^T \longrightarrow B := Q_1 \quad \longrightarrow \quad R_t \longrightarrow R_t (I_t \otimes M_1) R_t^T. \\ \text{for } B = Q_1 \end{array}$$

Using ADI we obtain in a canonical way that the composed system

$$R_t (I_t \otimes M_1) R_t^T \equiv Q_2 M_2 Q_2^T \tag{5.8}$$

is again a symmetric low-rank matrix. By construction, $Q_2 M_2 Q_2^T$ could be equivalently computed by applying t steps of the usual ADI method starting with initial guess $X_0 = 0$ and right hand side $-Q_1 M_1 Q_1^T$.

There are several structure-preserving Krylov subspace methods for (generalized) Lyapunov equations which are essentially based on the (block-) Krylov subspace

$$\text{span}\{B, AB, A^2 B, \dots, A^{k-1} B\},$$

see, e.g. [19–21, 23, 29, 41]. Krylov-subspace methods in conjunction with ADI preconditioning are frequently used [7, 17, 22], whereas the preservation of the low-rank structure of the iterates is not employed. Structure preservation of the GMRES and FGMRES methods [36] with LRCF-ADI preconditioning is further discussed in [9]. In [24] one can find a generalization of low-rank Krylov subspace methods for up to d-dimensional tensors.

5.3.3 Low-Rank Pseudo Arithmetic

The elementary matrix and vector operations preserve the symmetric low-rank format but numerically concatenation of symmetric low-rank matrices such as the linear combination may significantly increase the numerical rank of the iterates. To bypass this problem we need to introduce a pseudo arithmetic similar to the approach that is used for hierarchical matrices [15]. Let $Z = WMW^T$ with an additional inner small symmetric matrix $M \in \mathbb{R}^{l,l}$ be given. Z may have been obtained from one of the elementary operations described in Sect. 5.3.1. Then Z is compressed as follows:

1. We compute $W = QR\Pi^T$, where $Q \in \mathbb{R}^{n,r}$, $R \in \mathbb{R}^{r,l}$ and $\Pi \in \mathbb{R}^{l,l}$ using the QR decomposition with column pivoting [14]. To determine the rank using this QR decomposition has to be handled with care and should include the recent modifications suggested in [8], which is the case for LAPACK release 3.2 or higher. After truncation we obtain $W \approx Q_1 R_1 \Pi^T$.
2. Next we determine the eigenvalue decomposition $T = U\Sigma U^T$ of $T = R_1\Pi^T M\Pi R_1^T$ and reduce U, Σ to matrices U_1, Σ_1 of lower rank whenever the diagonal entries of Σ are sufficiently small in modulus.
3. This finally yields the truncated $W \approx (Q_1 U_1)\Sigma_1(Q_1 U_1)^T$, which is computed after each elementary operation, resp. after a sequence of elementary operations.

With respect to Krylov subspace methods we usually apply the iterative solver for solving $\mathscr{L}\mathscr{X} = \mathscr{B}$ until the norm of the residual $\|\mathscr{B} - \mathscr{L}\mathscr{X}_j\|_2 \leq \varepsilon$. Here ε may be an absolute or relative tolerance and may include contributions from $\mathscr{B}$. For generalized Lyapunov equations this condition reads as

$$\|EX_j A^T + AX_j E^T + P_l BB^T P_l^T\|_F \leq \varepsilon$$

and certainly any low-rank decomposition of R_j need not be significantly more accurate than ε. Whenever $EX_j A^T + AX_j E^T + P_l BB^T P_l^T \equiv W_j M_j W_j^T$ is compressed to lower rank, it is enough to compute a truncated $QR\Pi$ decomposition. To do so assume that

$$W_j = Q_j R_j \Pi_j^T$$

such that

$$R_j = \begin{pmatrix} R_{11} & R_{12} \\ 0 & R_{22} \end{pmatrix} = \left(\begin{array}{ccc|ccc} r_{11} & \cdots & r_{1p} & r_{1,p+1} & \cdots & r_{1,l} \\ & \ddots & \vdots & \vdots & & \vdots \\ 0 & & r_{pp} & r_{p,p+1} & \cdots & r_{p,l} \\ \hline & & & r_{p+1,p+1} & \cdots & r_{p+1,l} \\ & 0 & & \vdots & & \vdots \\ & & & r_{n,p+1} & \cdots & r_{nl} \end{array} \right).$$

The QR decomposition with column pivoting ensures that

$$|r_{11}| \geqslant \cdots \geqslant |r_{pp}| \geqslant \left\| \begin{pmatrix} r_{p+1,i} \\ \vdots \\ r_{n,i} \end{pmatrix} \right\|_2,$$

for all $i = p + 1, \ldots, l$. To make sure that the residual is accurate enough we may use a threshold tol_r, which should be chosen one order of magnitude less than ε and terminate the $QR\Pi$ decomposition as soon as

$$\max_{i=p+1,\ldots,l} \left\| \begin{pmatrix} r_{p+1,i} \\ \vdots \\ r_{n,i} \end{pmatrix} \right\|_2 \leqslant \mathrm{tol}_r . \tag{5.9}$$

This requires only a minor change to the $QR\Pi$ decomposition which is truncated as soon as the threshold is reached. Q_1, R_1 are then obtained by taking the first p columns of Q_j and the leading $p \times l$ block (R_{11}, R_{12}) of R_j multiplied by Π_j^T. In a similar way all other iterates of the low-rank Krylov subspace solver will be truncated to lower rank. To summarize our truncation strategy we give a small error analysis.

Lemma 5.3.1 *Let* $Z = WMW^T \in \mathbb{R}^{n,n}$ *such that* $W \in \mathbb{R}^{n,l}$, $M \in \mathbb{R}^{l,l}$ *for some* $l > 0$. *Suppose that the truncated* $QR\Pi$ *decomposition of* $W = QR\Pi^T$ *truncates the matrix* R *in* (5.9) *for some* $\mathrm{tol}_r = \varepsilon |r_{11}|$. *Discarding* R_{22}, *the approximate factorization*

$$\tilde{Z} = Q \begin{pmatrix} R_{11} & R_{12} \\ 0 & 0 \end{pmatrix} \Pi^T M \Pi \begin{pmatrix} R_{11} & R_{12} \\ 0 & 0 \end{pmatrix}^T Q^T$$

satisfies

$$\|Z - \tilde{Z}\|_2 \leqslant 2\sqrt{l-p}\, \varepsilon \|M\|_2 \|W\|_2^2 + \mathcal{O}(\varepsilon^2).$$

Moreover, suppose that

$$T := \begin{pmatrix} R_{11} & R_{12} \end{pmatrix} \Pi^T M \Pi \begin{pmatrix} R_{11} & R_{12} \end{pmatrix}^T$$

is decomposed as

$$T = U \Sigma U^T = (U_1, U_2) \begin{pmatrix} \Sigma_1 & 0 \\ 0 & \Sigma_2 \end{pmatrix} (U_1, U_2)^T$$

such that $U \in \mathbb{R}^{p,p}$ is orthogonal, $\Sigma_1 = \mathrm{diag}(\sigma_1, \ldots, \sigma_r)$, $\Sigma_2 = \mathrm{diag}(\sigma_{r+1}, \ldots, \sigma_p)$, $|\sigma_1| \geqslant \cdots \geqslant |\sigma_p|$ and $|\sigma_i| \leqslant \varepsilon|\sigma_1|$ for all $i > r$, then the approximate low rank factorization

$$\hat{Z} = \left(Q \begin{pmatrix} I_p \\ 0 \end{pmatrix} U_1 \right) \Sigma_1 \left(Q \begin{pmatrix} I_p \\ 0 \end{pmatrix} U_1 \right)^T$$

satisfies

$$\|Z - \hat{Z}\|_2 \leqslant (2\sqrt{l-p} + 1)\varepsilon \|M\|_2 \|W\|_2^2 + \mathcal{O}(\varepsilon^2).$$

Proof We first note that

$$|r_{11}| = \max_{j=1,\ldots,l} \|Re_j\| \leqslant \max_{\|x\|_2=1} \|Rx\|_2 = \|R\|_2 = \|W\|_2.$$

Conversely, using (5.9) we obtain

$$\|R_{22}\|_2 = \max_{\|y\|_2=1} \|R_{22}y\|_2 = \max_{\|y\|_2=1} \| \sum_{i>p} R_{22}e_i y_i \|_2$$

$$\leqslant \max_{\|y\|_2=1} \sum_{i=1}^{l-p} \|R_{22}e_i\|_2 \, |y_i|$$

$$\leqslant \max_{\|y\|_2=1} \left(\sum_{i=1}^{l-p} \|R_{22}e_i\|_2^2 \right)^{1/2} \left(\sum_{i=1}^{l-p} |y_i|^2 \right)^{1/2}$$

$$\leqslant \left((l-p)\varepsilon^2|r_{11}|^2 \right)^{1/2} \leqslant \sqrt{l-p}\, \varepsilon \|W\|_2.$$

It follows that

$$Z - \tilde{Z} = Q \begin{pmatrix} 0 & 0 \\ 0 & R_{22} \end{pmatrix} \Pi^T M W^T + W M \Pi \begin{pmatrix} 0 & 0 \\ 0 & R_{22} \end{pmatrix}^T Q^T$$

$$+ Q \begin{pmatrix} 0 & 0 \\ 0 & R_{22} \end{pmatrix} \Pi^T M \Pi \begin{pmatrix} 0 & 0 \\ 0 & R_{22} \end{pmatrix}^T Q^T.$$

Thus bounding the norm of $Z - \tilde{Z}$ yields

$$\|Z - \tilde{Z}\|_2 \leq 2\|R_{22}\|_2\|M\|_2\,\|W\|_2 + \|R_{22}\|_2^2\|M\|_2 \leq 2\sqrt{l-p}\,\varepsilon\|M\|_2\|W\|_2^2 + \mathscr{O}(\varepsilon^2).$$

Next observe that $\|T\|_2 = |\sigma_1|$ and we can bound $\|T\|_2$ by

$$\|T\|_2 \leq \|M\|_2\,\|\begin{pmatrix} R_{11} & R_{12} \end{pmatrix}\|_2^2 \leq \|M\|_2\,\|W\|_2^2.$$

If we now further truncate T, then

$$\|Z - \hat{Z}\|_2 \leq \|Z - \tilde{Z}\|_2 + \|\tilde{Z} - \hat{Z}\|_2$$

$$\leq 2\sqrt{l-p}\,\varepsilon\|M\|_2\|W\|_2^2 + \mathscr{O}(\varepsilon^2) + \|(Q\begin{pmatrix} I_p \\ 0 \end{pmatrix} U_2)\,\Sigma_2\,(Q\begin{pmatrix} I_p \\ 0 \end{pmatrix} U_2)^T\|_2$$

$$\leq 2\sqrt{l-p}\,\varepsilon\|M\|_2\|W\|_2^2 + \|\Sigma_2\|_2 + \mathscr{O}(\varepsilon^2)$$

$$\leq 2\sqrt{l-p}\,\varepsilon\|M\|_2\|W\|_2^2 + \varepsilon|\sigma_1| + \mathscr{O}(\varepsilon^2)$$

$$\leq (2\sqrt{l-p} + 1)\varepsilon\|M\|_2\|W\|_2^2 + \mathscr{O}(\varepsilon^2),$$

which completes the proof.

Although we may have $\|Z\|_2 < \|M\|_2\|W\|_2^2$ we consider this situation as rare in practice. Furthermore, the factor $\sqrt{l-p}$ is more of technical nature. Therefore using some $\tilde{\varepsilon}$ of one order of magnitude less than ε, we expect the truncation strategy to be in practice satisfactory in order to obtain $\|Z - \hat{Z}\|_2 \leq \varepsilon\|Z\|_2$. In Sect. 5.4 we will demonstrate the effectiveness of our approach.

To accommodate the preservation of symmetric low-rank matrices during elementary operations with the truncation to lower rank, a library LR-BLAS (**L**ow **R**ank-**B**asic **L**inear **A**lgebra **S**ubroutines) is designed which is summarized in Table 5.1.

The introduction of low-rank BLAS allows for the easy truncation to lower rank after an elementary operation is performed. We indicate and control whether only a concatenation of matrices is built or if rank compression is required. Even when the rank is to be reduced we can internally distinguish between only using the truncated $QR\Pi$ decomposition or reducing the rank further with the help of an eigenvalue decomposition. Also, we can handle the case when one of the symmetric low-rank

Table 5.1 Overview LR-BLAS library

Operation	Function reference
$\mathscr{Y} \leftarrow \mathscr{Y} + \alpha\mathscr{X}$	`lraxpy`
$\mathscr{Y} \leftarrow \alpha\mathscr{L}\mathscr{X} + \beta\mathscr{Y}$	`lrgemv`
$\mathscr{Y} \leftarrow \alpha\mathscr{Y}$	`lrscal`
$\alpha \leftarrow \|\mathscr{Y}\|$	`lrnorm`
$\alpha \leftarrow (\mathscr{Y}, \mathscr{X})$	`lrdot`

input matrices ($\mathscr{X}$ or $\mathscr{Y}$) already consists of orthonormal factors $X = QMQ^T$ such that $Q^T Q = I$. In this case one can simplify the amount of work when applying the QR decomposition. Internally, it is more convenient to represent a low-rank matrix $X = QRMR^T Q^T$ rather than $X = QMQ^T$. For the sequel of this article we will skip this detail.

The introduction of a low-rank pseudo arithmetic has immediate consequences when being used for generalized projected Lyapunov equations. While concatenation of symmetric low-rank matrices does not require any additional safe guard strategy, the situation changes as soon as the rank is compressed. After each rank compression with thresholds larger than the machine precision, the projectors P_l and P_r have to be applied again. In particular iterates such as the approximate solution $X_k \approx R_k M_k R_k^T$ require a projection step $X_k \to P_r R_k M_k R_k^T P_r^T = \hat{X}_k$ while iterates like the residual have to be treated differently. Recall that we have

$$E\hat{X}_k A^T + A\hat{X}_k E^T + P_l BB^T P_l^T = P_l(E\hat{X}_k A^T + A\hat{X}_k E^T + BB^T)P_l^T$$
$$\approx S_k N_k S_k^T,$$

thus here we obviously need to project with P_l to ensure that the iterates are mapped back to the correct invariant subspace associated with the finite eigenvalues of (E, A).

5.3.4 Approximate LRCF-ADI Preconditioning

Independent of the use of a low-rank pseudo arithmetic in Sect. 5.3.3, the explicit projection of the preconditioned iterate R_t from (5.8) gives the opportunity to replace the explicit inverses $(E + \tau_j A)^{-1}$ by an approximate inverse, e.g., using incomplete LU factorizations. Recall that when t steps of LRCF-ADI preconditioning are applied to a right hand side $B = P_l B$, then each iterate $R_j, j = 1, 2, \ldots, t$ satisfies $R_j = P_r R_j$. This is certainly not longer fulfilled when $(E + \tau_j A)^{-1}$ is replaced by an approximation. If in doubt, in any LRCF-ADI preconditioning step substitutes

$$(E + \tau_j A)^{-1} \to P_r \widetilde{(E + \tau_j A)}^{-1}$$

and explicitly projects the approximate solution back. In Sect. 5.4 we will demonstrate the effect of replacing the exact LU factorization of $E + \tau_j A$ by an ILU. At this point we like to stress that (low-rank) Krylov subspace methods are much less sensitive to the use of an ILU for $E + \tau_j A$ while the usual ADI method is much more affected.

5.3.5 Selected Low-Rank Krylov Subspace Methods

We now give some examples of preconditioned Krylov subspace methods adapted for generalized, projected Lyapunov equations using CFADI preconditioning. The most popular method, at least when E and A are symmetric and positive definite, is the conjugate gradient method. We will demonstrate the changes for this method first.

Suppose we wish to solve a system $\mathscr{L}\mathscr{X} = \mathscr{B}$ with a symmetric positive definite matrix $\mathscr{L}$ and a symmetric positive definite preconditioner $\tilde{\mathscr{L}} \approx \mathscr{L}$. Then the preconditioned CG method reads as given in Algorithm 5.2.

Now for symmetric and positive definite E and A we have $P_l = P_r$ and the generalized projected Lyapunov equation

$$EXA + AXE + P_l BB P_l^T = 0 \text{ where } X = P_r^T X P_r$$

induces the following preconditioned low-rank version Algorithm 5.3 with CFADI preconditioning and given shifts $\tau_1, \ldots, \tau_t$.

We will formally assume that each iterate Y is represented as $Y = Q_Y M_Y Q_Y^T$ for suitable matrices Q_Y and symmetric M_Y.

While the LR-BLAS internally apply rank compression and projection with P_l, for the preconditioning step one has to mention this explicitly to be consistent. A compression and projection step of P looks as follows.

$$P = R_t(I_t \otimes M_R)R_t^T \equiv Q_P M_P Q_P^T$$

by simple concatenation. Next the rank compression as described in Sect. 5.3.3 is performed and we obtain

$$(Q_P, M_P) \to (Q_P^{(new)}, M_P^{(new)}).$$

Algorithm 5.2 Preconditioned CG method

Let $\mathscr{X}_0 \in \mathbb{R}^n$ be initial guess
$\mathscr{R}_0 = -\mathscr{B} - \mathscr{L}\mathscr{X}_0$
$\mathscr{P} = \tilde{\mathscr{L}}^{-1}\mathscr{R}_0$
for $k = 1, 2, 3 \ldots$ **do**
    $\rho_{old} = \rho$
    $\mathscr{Z} = \mathscr{L}\mathscr{P}$
    $\alpha = (\mathscr{R}^T\mathscr{R})/(\mathscr{P}^T\mathscr{Z})$
    $\mathscr{X} = \mathscr{X} + \alpha\mathscr{P}$
    $\mathscr{R} = \mathscr{R} - \alpha\mathscr{Z}$
    $\mathscr{Z} = \tilde{\mathscr{L}}^{-1}\mathscr{R}$
    $\rho = \mathscr{R}^T\mathscr{Z}$
    $\beta = \rho/\rho_{old}$
    $\mathscr{P} = \mathscr{Z} + \beta\mathscr{P}$
end for

Algorithm 5.3 LR-CG for Lyapunov equations with CFADI preconditioning

---

$X_0 = 0, R_0 = -(P_l B)(P_l B)^T$
Compute $P = R_t(I_t \otimes M_{R_0})R_t^T$ using t steps of LRCF-ADI applied to $B = Q_{R_0}$
Compress and project P
$\rho = \text{trace}(RP)$ using `lrdot`
for $k = 1, 2, 3 \ldots$ **do**
    $\rho_{old} = \rho$
    $Z = EPA + APE$ using `lrgemv`
    $\alpha = \|R\|_F / \text{trace}(PZ)$ using `lrnorm` and `lrdot`
    $X = X + \alpha P$ using `lraxpy`
    $R = R - \alpha Z$ using `lraxpy`
    Compute $Z = R_t(I_t \otimes M_R)R_t^T$ using t steps of LRCF-ADI applied to $B = Q_R$
    Compress and project Z
    $\rho = \text{trace}(RZ)$ using `lrdot`
    $\beta = \rho / \rho_{old}$
    $P = Z + \beta P$ using `lrscal` and `lraxpy`
end for

Eventually P_l is applied, which yields

$$Q_P \to P_l Q_P \equiv Q_P^{(new)}.$$

One may or may not add another rank compression step to Q_P as a result of the projection. But this would have to be done accurately with respect to the machine precision.

The conjugate gradient method is designed for symmetric positive definite problems. This in turn only requires P_l. In general one has to distinguish which projection has to be applied. We demonstrate that in Algorithm 5.4 for the preconditioned GMRES method [37].

We point out that the use of LR-BLAS allows to only concatenate matrices or to compress the rank. Similarly, the projection need not always be applied. We have formulated the algorithms in this more general form to indicate which projection P_l or P_r is used. The basic operation $V^{(1)} = R/\rho$ usually does neither require rank compression nor projection. But if B would not have been projected before, a projection would be required at this point. Similarly, rank compression would usually not be used as long as B does not have a rank much less than the number of columns. For the preconditioning step using t steps of LRCF-ADI, formally there is no need to project W at the end, except if the rank were compressed. Numerically however, applying the projection may reduce the influence of rounding errors from previous preconditioning steps $j, j = 1, \ldots, t$.

Algorithm 5.4 LR-GMRES for Lyapunov equations with CFADI preconditioning

$X_0 = 0, R_0 = (P_l B)(P_l B)^T$
$\rho = \|R\|_F$ using `lrnorm`
$V^{(1)} = R/\rho$ using `lrscal(`P_l`)`
for $k = 1, 2, 3 \ldots, m$ **do**

Compute $W = R_t(I_t \otimes M_V^{(k)})R_t^T$ using t steps of LRCF-ADI applied to $B = Q_V^{(k)}$
Compress and project W by P_r
$Z = EWA^T + AWE^T$ using `lrgemv(`P_l`)`
for $l = 1, 2, 3 \ldots, k$ **do**

$h_{lk} = \text{trace}(V^{(l)}Z)$ using `lrdot`
$Z = Z + h_{lk}V^{(l)}$ using `lraxpy(`P_l`)`
end for

$h_{k+1,k} = \|Z\|_F$ using `lrnorm`
$V^{(k+1)} = Z/h_{k+1,k}$ using `lrscal(`P_l`)`
end for
Solve $\|\rho e_1 - \underline{H}_m y\|_2 = \min!$, where $\underline{H}_m = \left(h_{ij}\right)_{\substack{i=1,\ldots,m+1 \\ j=1,\ldots,m}}$
$Z = V^{(1)}y_1 + \cdots + V^{(m)}y_m$ using `lraxpy(`P_l`)`
Compute $W = R_t(I_t \otimes M_Z)R_t^T$ using t steps of LRCF-ADI applied to $B = Q_Z$
Compress and project W by P_r
$X = X + W$ using `lraxpy(`P_r`)`

The GMRES method can be slightly modified to obtain the flexible GMRES method (FGMRES, [35]). In this case, W would be replaced by $W^{(l)}$ and be kept. Then X is directly computed from $W^{(1)}, \ldots, W^{(m)}$ via

$$X = X + W^{(1)}y_1 + \cdots + W^{(m)}y_m \text{ using lraxpy(}P_r\text{)}.$$

FGMRES allows for variable preconditioning. This implies that the rank in $W^{(1)}, \ldots, W^{(m)}$ can be truncated with a larger tolerance tol_p than for the other iterates.

5.3.6 Reduced Lyapunov Equation

Several Arnoldi- and GMRES-like methods for Lyapunov equations essentially rely on the (block-) Krylov subspace $\text{span}\{B, AB, A^2B, \ldots, A^{k-1}B\}$ (see, e.g., [19–23]). These methods compute subspaces which replace the generalized Lyapunov equation (5.7) by a reduced equation

$$(WET) \tilde{X} (WAT)^T + (WAT) \tilde{X} (WET)^T + WP_l BB^T P_l^T W^T = 0.$$

The resulting approximate solution could be obtained from $X_k = P_r T \tilde{X} T^T P_r^T$. A similar approach would be possible as by product of the FGMRES method in order to obtain an alternative approximate solution. Suppose that the Arnoldi method applied to the Lyapunov operator $\mathscr{L}$ leads to the following equation

$$\mathscr{L}\mathscr{W}_m = \mathscr{V}_{m+1}\underline{\mathscr{H}}_m,$$

where $\mathscr{V}_m \in \mathbb{R}^{n^2,m}$ has orthonormal columns, $\underline{\mathscr{H}}_m \in \mathbb{R}^{m+1,m}$ is upper Hessenberg and the approximate FGMRES solution is given by $\mathscr{X}_m = \mathscr{X}_0 + \mathscr{W}_m s$ for $\mathscr{W}_m \in \mathbb{R}^{n^2,m}$. For the flexible GMRES method the columns of $\mathscr{W}_m$ are usually preconditioned counter parts of $\mathscr{V}_m$, except that the preconditioner may vary from step to step. Minimizing the norm of the residual $\mathscr{B} - \mathscr{L}\mathscr{X}_m$ for the standard GMRES method is equivalent to the minimization of

$$\|\underline{\mathscr{H}}_m y - \|\mathscr{R}_0\|_2 \cdot e_1\|_2 = \min! \tag{5.10}$$

Here one uses the property that the first column of $\mathscr{V}_m$ is chosen as a scalar multiple of the initial residual $\mathscr{R}_0 = \mathscr{B} - \mathscr{L}\mathscr{X}_0$. The Arnoldi vectors $\mathscr{V}_m e_k$ are rewritten in terms of symmetric low-rank matrices $V^{(k)} = Q_V^{(k)} M_V^{(k)} (Q_V^{(k)})^T$, $k = 1, \ldots, m$. Similarly, during the FGMRES method approximations to column k of $\mathscr{W}_k$ are represented by $W^{(k)} = Q_W^{(k)} M_W^{(k)} (Q_W^{(k)})^T$ from the CFADI preconditioning step. Then the numerical solution in low-rank format is a linear combination

$$X_k = X_0 + \sum_{k=1}^{m} y_k \, Q_W^{(k)} M_W^{(k)} (Q_W^{(k)})^T,$$

where the parameters $y = (y_1, \ldots, y_m)^T$ are taken from the minimization of the least squares problem (5.10). Alternatively the computed matrices $\left(Q_W^{(k)}\right)_k$ and $\left(Q_V^{(k)}\right)_k$ could be used to compute an alternative approximate solution $\hat{X}_k$.

Suppose that we compute a QR decomposition with column pivoting [14] to obtain

$$[Q_V^{(1)}, \ldots, Q_V^{(m)}] = Q_V R_V \Pi_V^T, \ [Q_W^{(1)}, \ldots, Q_W^{(m)}] = Q_W R_W \Pi_W^T,$$

where $\mathrm{rank} R_V = r_V$, $\mathrm{rank} R_W = r_W$. Similar to the compression to lower rank at other parts of the Krylov subspace method here one could work with lower accuracy as well. Let $r = \max\{r_V, r_W\}$, then the numerical solution X_k can be rewritten as

$$X_k = X_0 + Q_W S Q_W^T, \text{ where } S = R_W \Pi_W^T \begin{pmatrix} s_1 M_W^{(1)} & & 0 \\ & \ddots & \\ 0 & & s_m M_W^{(m)} \end{pmatrix} \Pi_W R_W^T.$$

Q_V and Q_W can be alternatively used to construct a reduced r-dimensional Lyapunov equation. Let

$$E_Q = Q_V^T E Q_W, A_Q = Q_V^T A Q_W$$

and compute S as numerical solution of the reduced equation

$$E_Q \, S \, A_Q^T + A_Q \, S \, E_Q^T + Q_K^T R_0 Q_K = 0,$$

where $R_0 = EX_0A^T + AX_0E^T + BB^T$. For small r this could be computed with standard methods [2]. We obtain

$$\hat{X}_m = X_0 + Q_W S Q_W^T$$

as approximate solution of a reduced Lyapunov equation. In Sect. 5.4 we will demonstrate the effectiveness of this approach.

In summary the low-rank Krylov subspace methods introduced in Sect. 5.3 allow for structured iterative methods. If $(P_l E P_r, P_l A P_r)$ is already symmetric and $P_l E P_r$ positive semidefinite, one could use a low-rank version of the simplified QMR (SQMR) method [12] for symmetric indefinite problems. If even $P_l A P_r$ is positive definite, then the low-rank CG method can be applied. Low-rank CG and low-rank SQMR can make use of the CFADI preconditioning approach while at the same time low-rank structures and symmetry of the Lyapunov operator is preserved. In the general case we could easily introduce low-rank Krylov subspace methods such as low-rank BiCGStab, low-rank QMR and other methods (cf. [36]).

5.4 Numerical Results

In this section we will demonstrate the effectiveness of our approach. We will start with the sensitivity of low-rank Krylov subspace methods with respect to the shifts used for the CFADI preconditioning step and compare them with the usual LRCF-ADI method. Next we will demonstrate different low-rank Krylov subspace methods such as (F)GMRES, QMR and BICGSTAB for projected, generalized Lyapunov equations to evaluate their strengths and their weaknesses. We will further investigate replacing the direct solver for the single iterates $(E + \tau_j A)^{-1}$ by an approximate factorization to compare the sensitivity of ADI and Krylov subspace methods with respect to incomplete factorizations. Here we use as approximate factorization the multilevel ILU factorization from the software package[1] ILUPACK which is described in detail in [5]. Further numerical results will discuss the use of the reduced equation from Sect. 5.3.6 for the numerical solution. We will finally demonstrate how parallel direct solvers can accelerate the process of solving large-scale projected Lyapunov equations.

---

[1]Matthias Bollhöfer and Yousef Saad. ILUPACK - preconditioning software package. Available online at http://ilupack.tu-bs.de/.ReleaseV2.4,June2011.

Some of our experiments use the software package PABTEC, see [33] and Sect. 2.6, which has been designed for the model order reduction of descriptor systems arising from circuit simulation. Here we replaced the default LRCF-ADI method by preconditioned low-rank Krylov subspace methods such as (F)GMRES, QMR and BICGSTAB and adapted the interfaces to allow for complete simulation runs based on Krylov subspace techniques.

5.4.1 Model Problems

In the following part we like to introduce three model problems which we will use for demonstration. The first two are examples arise from descriptor systems modeling circuit-equations while the third one is a more academic parabolic partial differential equation. All these examples illustrate the applicability of low-rank Krylov subspace methods.

As our first two examples we discuss linear RLC networks of the following type, modeled using the modified nodal analysis (MNA). Let e be the vector of node potentials, v_V, v_I be the voltages of the voltage sources, respectively of the current sources. Denote by i_L, i_V, i_I the currents through the inductors, voltage sources and current sources. We define the state vector x, the vector of inputs u and the output vector y via

$$x = \begin{pmatrix} e \\ i_L \\ i_V \end{pmatrix}, \quad u = \begin{pmatrix} i_I \\ v_V \end{pmatrix}, \quad y = \begin{pmatrix} v_I \\ i_V \end{pmatrix}.$$

Then the circuit equations can be written as

$$E\dot{x} = Ax + Bu$$
$$y = -B^T x,$$

where E, A and B are given by

$$E = \begin{pmatrix} A_C C A_C^T & 0 & 0 \\ 0 & L & 0 \\ 0 & 0 & 0 \end{pmatrix}, \quad A = \begin{pmatrix} -A_R G A_R^T & -A_L & -A_V \\ A_L^T & 0 & 0 \\ A_V^T & 0 & 0 \end{pmatrix}, \quad B = \begin{pmatrix} -A_I & 0 \\ 0 & 0 \\ 0 & -I \end{pmatrix}.$$

Here A_C, A_R, A_L, A_V, A_I refer to the incidence matrices with respect to the capacitors, resistors, inductors, as well as with respect to the voltage sources and current sources. C, L, G denote the capacitance matrix, the inductance matrix and the conductivity matrix. The differential-algebraic equations which we discuss here are of differentiation index 1 (cf. [6]).

Table 5.2 Large-scale RC circuits

Acronym	Capacitors	Resistors	Voltage sources	System size
RC1	2353	1393	109	974
RC2	3065	5892	21	3272
RC3	9999	9999	3	10, 002
RC4	12, 025	53, 285	78	29, 961

Example 5.4.1 As a first example we consider a RC high pass circuit provided by NEC Laboratories Europe. It consists of 2002 conductors, 2003 resistors and three voltage sources. Using the MNA this leads to a system of dimension 2007 with three inputs and three outputs.

Example 5.4.2 We consider further test[2] examples of several RC circuits. For some details we refer to [18]. Here we restrict ourselves to examples of following sizes, reported in Table 5.2.

The circuits in Table 5.2 are of differentiation index 2. Since we like to demonstrate the applicability of low-rank Krylov subspace methods for index-1 systems we remove several voltage sources which are responsible for the higher index. After removing these voltage sources we have for circuit RC1, six voltage sources and for each circuit RC2, RC3 and RC4, one voltage source. Furthermore, we artificially add resistors with average conductivity to $A_R G A_R^T$ to make this matrix positive definite. We are aware of changing the original shape of these circuits. However, our main goal is the demonstration of low-rank Krylov subspace methods using the **PABTEC** software as framework.

For both problem classes of RC circuits in Examples 5.4.1 and 5.4.2 we use the technology as provided by the software package **PABTEC** (see Sect. 2.6 and [33]) to demonstrate solving an associated projected algebraic Riccati equation with the help of Newton's method. Here in every Newton iteration step (cf. Algorithm 2.7 in Chap. 2) a projected, generalized Lyapunov equation has to be solved.

Example 5.4.3 The final example we will use in our numerical experiments is the parabolic partial differential equation

$$v_t = \Delta v + \mathscr{B}u \equiv v_{xx} + v_{yy} + v_{zz} + \mathscr{B}u,$$

where $v = v(x, y, z, t)$, $(x, y, z) \in \Omega = [0, 1]^3$ and $t \geq 0$. We assume that we have some initial value $v(x, y, z, 0)$ and homogeneous Dirichlet boundary conditions. To keep the discussion simple, we consider an academic control $\mathscr{B}$ such that after discretization in space using a seven-point discretization stencil, the control reduces to the vector with all ones. Suppose that we have an equidistant mesh with mesh size $h = \frac{1}{N+1}$. This leads to a total system size of $n = N^3$ unknowns. The

---

[2]http://sites.google.com/site/rionutiu2/research/software.

semi-discretized ordinary differential equation is of type

$$\dot{w} = -Aw + Bu,$$

where A is the discretized Laplacian operator in three spatial dimensions. We apply model order reduction to these semi-discretized equations using balanced truncation. For symmetry reasons we simply compute the associated Gramian as the solution of the Lyapunov equation

$$XA + AX = BB^T,$$

meaning N^6 unknowns for the referring Lyapunov operator. Since A is symmetric and positive definite, the Lyapunov equation $X(-A) + (-A)X + BB^T = 0$ is stable and therefore balanced truncation can be applied. We know that the spectrum of A lies inside the interval $(3\pi^2, \frac{12}{h^2})$. This allows for a simple computation of the optimal ADI shift-parameters introduced by Wachspress [47].

We use this example in order to illustrate a low-rank version of the conjugate gradient method. Furthermore, a parallel sparse direct solver for solving the shifted systems $(A + \tau_i I)x = b$ is used to examine the scalability. Finally, this example demonstrates the advantages of using multilevel incomplete factorizations rather than direct solvers within the CFADI method.

In the sequel all computations were conducted on a 64 GB Linux workstation with four Intel Xeon E7440 Quadcore processors using Matlab Release R2008b.

5.4.2 Different Krylov Subspace Methods and Their Efficiency with Respect to the Selection of Shifts

In the following experiments we will compare how flexible GMRES [35], GMRES [37], QMR [13] and BICGSTAB [46] can be used to solve projected generalized Lyapunov equations. We will describe how different choices of shifts affect the LRCF-ADI method and low-rank Krylov subspace methods. For this purpose we consider Examples 5.4.1 and 5.4.2. Here it is necessary to use the heuristic approach (referred to as *"Algorithm 1"* in [32]) for calculating the shift parameters. As part of the passivity-preserving balanced truncation we will solve the projected Riccati equations from (5.4), (5.5) up to a tolerance of 10^{-4}. The same accuracy is used for truncating the Hankel singular values for Balanced Truncation. As a heuristic approach we decided to solve each Lyapunov equation up a relative residual norm of 10^{-6}. One benefit of our class of Krylov subspace methods is that we can use the norm provided by our Krylov-subspace method and do not need to explicitly evaluate the residual-norm within the LRCF-ADI algorithm. We vary the number t of calculated shift parameters from 4, 5, 10, 20 finally to 30. For the low-rank Krylov methods we use a tolerance of 10^{-8} for truncating the ranks which is two

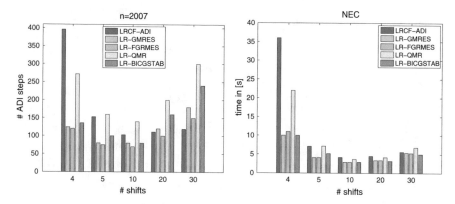

Fig. 5.1 Number of ADI steps and runtime for Example 5.4.1

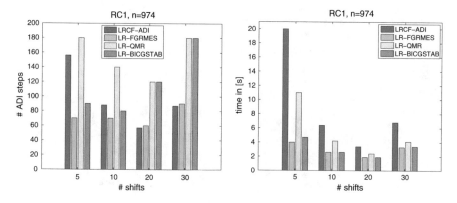

Fig. 5.2 Number of ADI steps and runtime for circuit RC1 from Example 5.4.2

orders of magnitude smaller than the desired residual. The number of ADI steps we display in Figs. 5.1, 5.2, 5.3, 5.4 and 5.5 refer to the accumulated sum of all shifted systems that were solved using Newton's method.

As can be seen from Figs. 5.1, 5.2, 5.3, 5.4, and 5.5, there is neither a method that is always fastest nor is there a method always requiring the smallest number of ADI solving steps. Comparing flexible GMRES with standard GMRES, the difference in the number of ADI iterations can be explained by the different nature of these approaches. While the number of Krylov subspace iteration steps is the same, standard GMRES requires one additional solving step at the end of each restart. In contrast to this, flexible GMRES stores the preconditioned residuals explicitly and does not require an additional preconditioning step. The slightly improved computation time of flexible GMRES with respect to GMRES is obtained by using twice as many vectors in low-rank format. When working with restarts this is an acceptable tradeoff so we prefer to use flexible GMRES over standard GMRES in low-rank arithmetic.

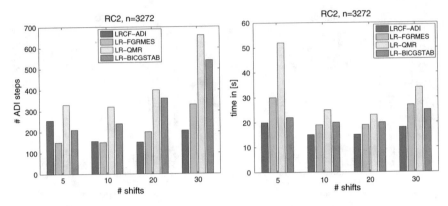

Fig. 5.3 Number of ADI steps and runtime for circuit RC2 from Example 5.4.2

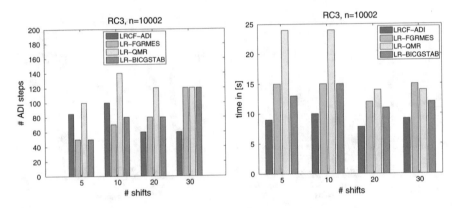

Fig. 5.4 Number of ADI steps and runtime for circuit RC3 from Example 5.4.2

BICGSTAB and QMR require in each iteration step that either the matrix is applied twice (BICGSTAB) or the transposed matrix is used in addition (QMR). The same holds for the application of the preconditioner. Often BICGSTAB is comparable to GMRES with respect to time while QMR is typically the slowest method.

We emphasize that the number of inner iteration steps for the projected Lyapunov equations is small, when a larger number of shifts is used. When using $t = 20$ or $t = 30$ shifts, the number of inner iteration steps is typically less than ten steps. We illustrate the relation between inner ADI solving steps and outer Newton steps in Fig. 5.6 for the case of the LRCF-ADI method and LR-FGMRES and different numbers of shifts for Example 5.4.1.

The two graphics at the top of Fig. 5.6 refer to the use of four shifts while the two graphics at the bottom of Fig. 5.6 refer to the use of ten shifts. On the left of Fig. 5.6 we find the LRCF-ADI method, on the right LR-FGMRES is displayed.

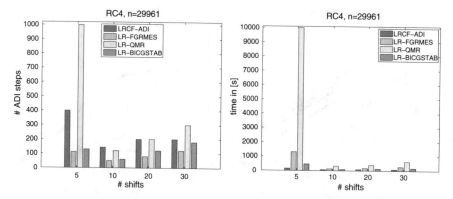

Fig. 5.5 Number of ADI steps and runtime for circuit RC4 from Example 5.4.2

The meaning of the three-coded numbers of type a-b-c in Fig. 5.6 is explained in the legend just below the graphics.

The solid red line in Fig. 5.6 reveals the norm of the nonlinear residual in Newton's method. The other lines display the convergence history of the residuals during the inner solves. In particular we observe that both methods, LRCF-ADI and LR-FGMRES reach the threshold 10^{-4} of the nonlinear residual after four outer steps. It can also be observed that LRCF-ADI using four shifts exceeds the limit 100 of inner iteration steps for solving the projected Lyapunov equation without converging. In spite of misconvergence, the outer Newton method in this case still converged to the desired accuracy.

5.4.3 Truncated $QR\Pi$ Decomposition

In this section we will demonstrate the difference in using the regular QR decomposition with column pivoting as implemented in LAPACK (also used inside MATLAB) with a truncated version that stops the decomposition as soon as the desired accuracy for the truncation is reached (for details cf. Sect. 5.3.3).

In Example 5.4.1 the main time for performing the Balanced Truncation algorithm is consumed when solving the Riccati equation. In Table 5.3 the computation time of the LR-FGMRES method using **PABTEC** for different numbers of shifts using the full $QR\Pi$ decomposition versus the truncated $QR\Pi$ is stated.

As solver for the Lyapunov equation we use LR-FGMRES. As in Sect. 5.4.2 both relative rank tolerances were set to 10^{-8} whereas we are solving the Lyapunov equations with accuracy 10^{-6}. The gain observed for using the truncated $QR\Pi$ was approximately in the range of about 5–8% in overall runtime of the LR-FGMRES method.

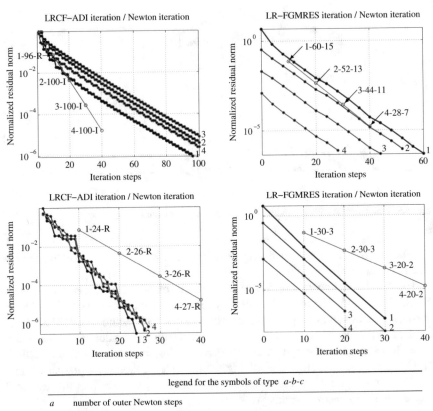

Fig. 5.6 Comparison of LRCF-ADI and LR-FGMRES using four (*top line*) and ten (*bottom line*) shifts

Table 5.3 Comparison of standard $QR\Pi$ and truncated $QR\Pi$ within LR-FGMRES, Example 5.4.1

# Shifts	Standard $QR\Pi$ (s)	Truncated $QR\Pi$ (s)
4	19.25	18.43
5	7.33	6.90
10	4.03	3.90
20	3.95	3.79
30	4.24	4.20

The improvement using the truncated $QR\Pi$ decomposition can not only be used in low-rank Krylov subspace methods, but it can also have a beneficial impact of the LRCF-ADI method. When solving the projected Riccati equations using LRCF-ADI, at each Newton step we have to concatenate the current approximate low-rank solution $Z = Q_Z Q_Z^T$ of the Riccati equation and the recent low-rank update $P = Q_P Q_P^T$ from solving the projected Lyapunov equation to obtain

$$Z + P = \begin{bmatrix} Q_Z & Q_P \end{bmatrix} \begin{bmatrix} Q_Z & Q_P \end{bmatrix}^T \underset{\text{compression}}{\overset{\text{rank}}{\longrightarrow}} Q_Z^{(new)} (Q_Z^{(new)})^T.$$

Usually we would apply a slim QR decomposition

$$\begin{bmatrix} Q_Z & Q_P \end{bmatrix} \overset{!}{=} QR$$

such that Q has as many columns as $\begin{bmatrix} Q_Z & Q_P \end{bmatrix}$. After that we would apply a singular value decomposition

$$R \overset{!}{=} U_r \Sigma_r V_r^T$$

to truncate the rank of R to some r and obtain

$$Q_Z^{(new)} = QU_r \Sigma_r.$$

When we use the truncated $QR\Pi$ decomposition instead, we can already compute approximately

$$\begin{bmatrix} Q_Z & Q_P \end{bmatrix} \overset{!}{=} Q_s R_s \Pi^T + E$$

such that $\|E\|$ is small and Q_s and R_s^T may already have significantly less columns s than $\begin{bmatrix} Q_Z & Q_P \end{bmatrix}$. Next a singular value decomposition only needs to be applied to the already reduced system

$$R_s \overset{!}{=} U_r \Sigma_r V_r^T.$$

Thus, the truncated $QR\Pi$ decomposition may not only save time during the $QR\Pi$ decomposition of $\begin{bmatrix} Q_Z & Q_P \end{bmatrix}$, but the singular value decomposition is also applied to system of smaller size and may lead to additional improvements. To illustrate this effect we compare the LRCF-ADI method for Examples 5.4.1 and 5.4.2. Although the total computation time is not drastically improved, at least the time of the rank compression is moderately improved. In Figs. 5.7 and 5.8 we illustrate the computation times of both rank compression techniques, accumulated over all Newton steps.

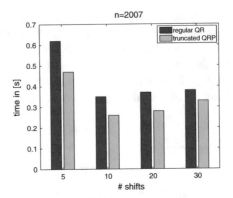

Fig. 5.7 Computation time QR plus SVD version truncated $QR\Pi$ plus SVD for Example 5.4.1

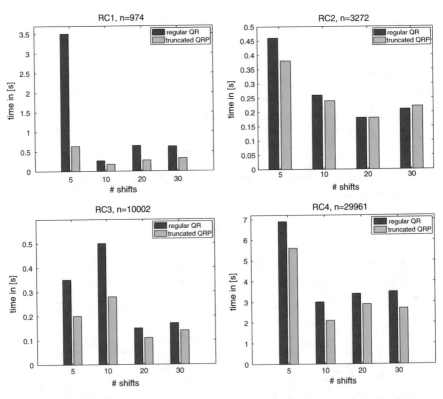

Fig. 5.8 Computation time regular QR plus SVD implementation versus truncated $QR\Pi$ plus SVD for Example 5.4.2

We can observe a moderate to significant gain in particular when using a smaller number of shifts. When only using a smaller number of shifts, the total number of ADI steps significantly increases since the LRCF-ADI method needs more steps to converge. This in turn results in a higher pseudo rank caused by simple concatenation. Here the gain is most significant.

5.4.4   Evolution of the Rank Representations in the Low-Rank CG Method

We will now report for the preconditioned LR-CG method from Algorithm 5.3 how ranks of the symmetric low-rank matrices X, R and P behave during the iterative process. To illustrate their behaviour we select Example 5.4.3 since we believe that the LR-CG method is the easiest low-rank Krylov subspace method and this example allows for the use of the preconditioned LR-CG method. We select as discretization parameter $N = 60$ which lead to a sparse symmetric positive definite matrix A of size $n = N^3 = 216,000$. The associated Lyapunov equation $X(-A) + (-A)X + BB^T = 0$ is numerically solved to obtain a low-rank symmetric positive semidefinite solution $X \in \mathbb{R}^{n,n}$. In the experiment we use a residual norm of 10^{-6} as termination criterion for the preconditioned LR-CG method. Since A is symmetric and positive definite we are able to use the optimal Wachspress shifts [47] for CFADI preconditioning. We demonstrate the behaviour of the ranks of X, R and P when using $t = 4, 6, 8$ and $t = 10$ shifts. For any of these shift values the LR-CG method only requires a few steps to converge (see Table 5.4).

In Fig. 5.9 we illustrate the behaviour of the ranks of X, R and P in the LR-CG method, when we use a truncation tolerance of 10^{-8}.

The solid lines in Fig. 5.9 refer to the situation where X, R and P are updated and truncated to lower rank in the LR-CG method, i.e., whenever the operations

$$X = X + \alpha P \text{ using } \texttt{lraxpy}$$
$$R = R - \alpha Z \text{ using } \texttt{lraxpy}$$
$$\ldots$$
$$P = Z + \beta P \text{ using } \texttt{lrscal} \text{ and } \texttt{lraxpy}$$

are completed within Algorithm 5.3. For X the dashed lines indicate the intermediate rank before the $\texttt{lraxpy}$ routine compresses the rank. Similarly, for R the dashed line indicates the pseudo rank before and after the rank truncation of Z in the

Table 5.4 Number of shifts and number of preconditioned LR-CG steps for Example 5.4.3 and $N = 60$

Number of shifts	4	6	8	10
Number of LR-CG steps	7	5	4	3

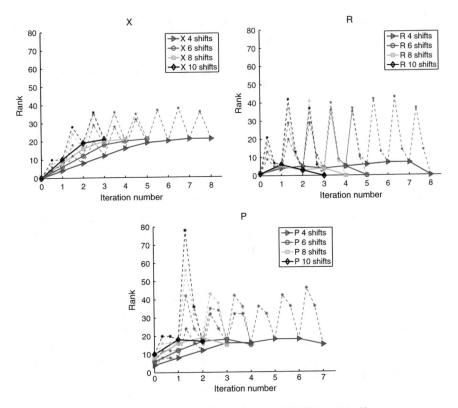

Fig. 5.9 Evolution of ranks for different selected vectors in LR-CG for $N = 60$

`lrgemv` routine that computes $Z = XA + AX$ and the situation before and after `lraxpy` compresses the rank for $R = R - \alpha Z$. Finally, the dashed line that is used for P includes the pseudo rank from the CFADI preconditioning step followed by its rank compression, as well as the additional rank compression, when $P = Z + \beta P$ is computed. We can observe for X, R and P that the intermediate ranks can be significantly higher than the rank that is obtained when `lraxpy` is completed. As we would expect, at the end of each rank compression step, the rank of X and P tends towards a constant rank, while the R the rank of the residual becomes small or even 0 when the LR-CG method converges. The general behaviour of the ranks, in particular that ranks first increase and then decrease again has also been observed in other low-rank Krylov subspace methods and applications [25]. The intermediate increase of the rank can be interpreted as another justification for using the truncated $QR\Pi$ decomposition to improve the performance of low-rank Krylov subspace methods as already illustrated in Sect. 5.4.3.

Fig. 5.10 Norm of the
residuals for LR-FGMRES
using different number of
shift parameters. Comparison
of usual LR-FGMRES versus
approximation via the
reduced system

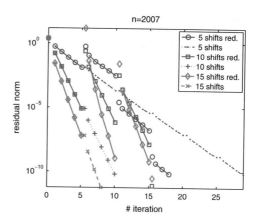

5.4.5 Numerical Solution Based on Reduced Lyapunov Equations

The LR-FGMRES method computes orthonormal Arnoldi vectors that can be
used to define a reduced projected Lyapunov equation (see Sect. 5.3.6). However,
although having this reduced Lyapunov equation available, the additional information we can extract from solving this reduced equation does not necessarily improve
the low-rank solution computed via LR-GMRES. To illustrate this effect we will
consider Example 5.4.1 using different number of shift parameters. Here we simply
examine solving a simple Lyapunov equation using a tolerance of 10^{-10} for the
residual and a truncation threshold for the rank of 10^{-12}.

The results are shown in Fig. 5.10, where the norm of the residual at the end of
m steps LR-FGMRES is compared with the version that uses the information of the
reduced system instead.

As we can see from Fig. 5.10 using the approximate solution from the reduced
system does not necessarily improve the residual. Moreover, the computational
overhead should not be overlooked. Solving the reduced system requires to solve a
small projected generalized Lyapunov equation using a method such as the Bartels-
Stewart algorithm. This increases the computational amount of work. For further
details we refer to [9].

5.4.6 Incomplete LU Versus LU

We now examine numerically how replacing the direct solver for $(E + \tau_j A)^{-1}$ by
the multilevel ILU from ILUPACK influences the LRCF-ADI method and the LR-
FGMRES method with LRCF-ADI preconditioning. First we use Example 5.4.1 to
compare both methods inside the model order reduction software package PABTEC.
Both iterative methods replace the direct solver by the ILU with a default threshold

Fig. 5.11 Norm of the
residuals for LRCF-ADI and
LR-FGMRES, both using
incomplete CFADI
preconditioning with 15 shifts

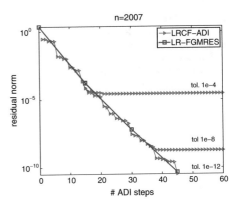

of 10^{-2} for discarding small entries. In addition, in our experiments the iterative solver inside ILUPACK [which is by default GMRES(30)] uses as termination criterion a relative residual of 10^{-4}, 10^{-8} and 10^{-12} to illustrate different accuracy of the multilevel ILU solver.

The results in Fig. 5.11 demonstrate that in principle low-rank Krylov subspace methods can use approximate factorizations rather than direct factorization methods while the usual LRCF-ADI method encounters convergence problems which are caused by solving $(E + \tau_i A)x = b$ with lower relative accuracy.

The convergence for the results in Fig. 5.11 is slightly delayed for LR-FGMRES while LRCF-ADI does not converge anymore. A drawback of the use of approximate factorizations that we observed in the numerical experiments is that the rank of the single iterates significantly increases [10]. This reduces the advantages of incomplete factorizations at least for these kind of examples where direct solvers are a natural alternative. The source of this increase will be subject to future research.

As second example we consider Example 5.4.3 where direct solvers quickly reach their limit because of the complexity and the spatial dimension. Besides, the Lyapunov equations in this case can be numerically solved using the preconditioned LR-CG method. Firstly we will compare the memory consumption. For the comparison we will use MATLAB's `chol` function that computes a Cholesky decomposition in combination with `symamd` which initially reorders the system using the symmetric approximate minimum degree algorithm [1] in order to save fill-in. In the sequel we will refer to this version as "MATLAB". Next we use for comparison the software package[3] PARDISO [39, 40] and its Cholesky decomposition. For the incomplete factorization we will again use ILUPACK and its inverse-based multilevel incomplete Cholesky factorization with the additional option to preserve the vector with all entries equal to 1 exactly. The latter is recommended since the underlying matrix refers to a discretized elliptic partial differential equation. Since the matrix is symmetric positive definite we again

---

[3] http://www.pardiso-project.org.

Table 5.5 Number of ADI shifts depending on N and tol_w

N	20			40			60			80		100
tol_w	10^{-1}	10^{-2}	10^{-4}	10^{-1}	10^{-2}	10^{-4}	10^{-1}	10^{-2}	10^{-4}	10^{-1}	10^{-2}	10^{-1}
Shifts	3	4	8	3	5	9	4	6	10	4	6	4

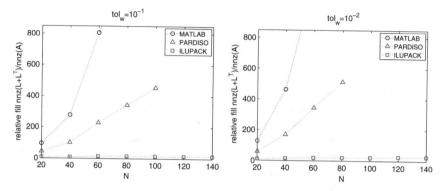

Fig. 5.12 Memory requirement illustrated by the relative fill-in of the Cholesky factor with respect to the given matrix

use the Wachspress shifts, similar to Sect. 5.4.4. Depending on the discretization parameter N these shifts are computed with respect to a given tolerance tol_w for the desired accuracy of CFADI approximation (for details we refer to the parameter ε_1 in [28]). In Table 5.5 we give the explicit relation between the number of shifts depending on N and tol_w.

In Fig. 5.12 we display how the relative fill-in $\frac{nnz(L+L^T)}{nnz(A)}$ of the nonzero entries of the Cholesky factor L relative to the nonzero entries of A behaves with respect to the discretization size N for $\mathrm{tol}_w = 10^{-1}$ and $\mathrm{tol}_w = 10^{-2}$.

As is well-known for problems in three spatial dimensions, the relative fill-in of direct solvers drastically increases when the size of the problem increases with PARDISO being significantly better than MATLAB. In contrast to that ILUPACK yields an almost constant relative fill-in for each tol_w and also only mildly increases when tol_w is decreased (i.e., when the number of shifts is increased). The increase in the amount of fill-in is significantly sublinear! We illustrate this effect for $N = 60$. Since we need to factorize $F_i = A + \tau_i I$, for $i = 1, 2, \ldots, t$ for each shift τ_i, the system F_i is almost equivalent to A, as long as a relatively small shift τ_i is chosen. Increasing the shift τ_i in magnitude, as it is happening in the computation of the optimal ADI shift parameters, makes F_i more and more diagonal dominant. When F_i is almost equivalent to A, the multilevel ILU requires more fill-in and more levels, since in this case a multigrid-like approximation is required. With increasing diagonal dominance of F_i, the multilevel ILU gets sparser and requires less fill-in, adapting automatically to the underlying system. This explains why even increasing the number of shifts does not necessarily result in a linear increase of memory or

Table 5.6 Performance of ILUPACK's multilevel ILU when four optimal shifts are prescribed, $N = 60$, $tol_w = 10^{-1}$

Shift value	t_{factor} (s)	Levels	Fill-in	t_{solve} (s)	Steps
$-26,999.996$	2.0	1	2.2	0.8	8
-3406.818	2.5	2	3.6	1.5	10
-387.730	5.2	3	3.9	1.6	13
-48.923	6.2	5	4.7	2.1	18

computation time. In Table 5.6 we state the computation time for computing the multilevel ILU for a system of size $N^3 = 216,000$ depending on the value of the shift.

We have chosen $tol_w = 10^{-1}$, which gives four shifts $\tau_1, \ldots, \tau_4$. For a large absolute value of $\tau_1 = -26,999.996$ the system is strictly diagonal dominant. Thus only 1 level is required, the computation time is small and the relative fill-in is approximately twice as much as that of the original system. With such a large shift, solving a single system with the multilevel ILU is not only fast because of the sparse approximation, but it also requires the fewest number of iteration steps (in this case 8 steps of preconditioned CG for a single right hand side). When the shift decreases in magnitude, the diagonal dominance becomes less, the number of levels increases and ILUPACK's multilevel ILU behaves more and more like an algebraic multilevel method. This can be verified by the increasing number of levels, the increasing fill-in and the slightly increasing number of CG steps.

The sublinear behaviour of the multilevel ILU is also a significant advantage with respect to the computation time when solving the Lyapunov equations using LR-CG with CFADI preconditioning. We state the computation time in Table 5.7.

As we can see from Table 5.7, the computation time behaves differently for different solvers when increasing the number shifts. Using more shifts result frequently in working with higher ranks also already seen in Fig. 5.9. This is because increasing the number of shifts only mildly increases the fill-in while at the same time the convergence speed is improved. Here ILUPACK is by far the fastest numerical solver for computing systems with $F_i = A + \tau_i I$. Looking at Table 5.7 we can also see that the computation time of the direct solvers scales significantly better than their memory requirement which is cause by sparse elimination technologies, such as the elimination tree, super nodes, Level-3-BLAS and cache optimization. These are techniques that are hardly applicable to incomplete factorization techniques.

5.4.7 Parallel Approach

We finally illustrate how the computation can be reduced for large-scale examples when the direct solver is replaced by a multi-threaded direct solver which can make use of several cores during the factorization and the solution phase. Here we use the direct solver PARDISO [39, 40] and demonstrate the different computation times when using several threads. For this purpose we again chose Example 5.4.3 since

Table 5.7 Computation time LR-CG in (s) using CFADI preconditioning with different inner solvers (MATLAB/PARDISO 1 cpu/ILUPACK)

Dimension N	# Shifts	MATLAB	PARDISO(1)	ILUPACK
20	3	8.5	3.5	3.5
	4	9.3	3.8	3.1
	8	12.8	4.7	2.8
40	3	596.6	144.9	75.8
	5	575.1	137.0	59.3
	9	673.8	156.6	48.4
60	4	9236.6	1564.4	375.8
	6	10,847.2	1717.4	284.3
	10	11,273.8	1879.9	271.2
80	4	78,870.4	10,562.7	1312.7
	6	–	10,475.9	1137.9
100	4	–	43,255.3	3653.7
100	6	–	–	2647.5
120	4	–	–	7551.8
120	7	–	–	6232.5
140	4	–	–	15,311.4
140	7	–	–	10,201.0

here we are able to adjust/increase the dimension of the equation. As solver we use the LR-CG method since we know in this case the equivalent linear system would be symmetric and positive definite. We increase the size of the matrix A from $20^3 = 8000$ to $100^3 = 1,000,000$. Remember that the corresponding Lyapunov equation would even have squared size. We will solve the Lyapunov equation up to a residual norm of 10^{-6}. For this example optimal shift parameters can be computed [47]. The number of shifts are computed according to a tolerance tol_w which refers to the convergence speed of the ADI method. Here we choose $tol_w = 10^{-1}$, $tol_w = 10^{-2}$ and $tol_w = 10^{-4}$ as tolerances. The number of shifts can be seen in the second column of Table 5.8.

The values are always ordered from $tol_w = 10^{-1}$ down to $tol_w = 10^{-4}$ (cf. also Table 5.5). For $N \geq 80$ we skipped $tol_w = 10^{-4}$ and for $N \geq 100$ we skipped $tol_w = 10^{-2}$ additionally for reasons of memory consumption.

Beside the computation time in Table 5.8 we point out that the number of LR-CG steps only depends on the size of tol_w. Numerically it is advantageous to have a larger value of tol_w and to use more LR-CG steps since this significantly saves memory and occasionally is even the fastest version as can be seen from Table 5.8.

Using the multithreaded parallel solver PARDISO we observe a significant speedup which is close to linear for larger N. It can also be seen that using 4 threads or 8 threads leads to an optimal performance on our machine. We observed that for maximum possible number of 16 threads the amount of computational time increased drastically. We blame this issue to problems of the dense linear algebra

190 M. Bollhöfer and A.K. Eppler

Table 5.8 Computation time LR-CG in (s) using CFADI preconditioning with a multithreaded version of PARDISO

Dimension N	# Shifts	cpu = 1	cpu = 2	cpu = 4	cpu = 8
20	3	3.5	3.2	3.2	8.9
	4	3.8	3.4	3.4	10.2
	8	4.7	4.1	4.1	11.8
40	3	144.9	87.5	77.2	124.0
	5	137.0	79.4	66.3	118.6
	9	156.6	88.0	73.5	131.4
60	4	1564.4	704.2	464.4	983.5
	6	1717.4	735.4	504.2	1064.3
	10	1879.9	794.8	622.8	1160.1
80	4	10,562.7	4121.1	2585.0	6448.1
	6	10,475.9	4032.2	2702.0	6432.6
100	4	43,255.3	15,363.7	9767.2	24,577.4

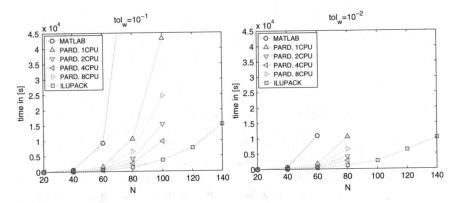

Fig. 5.13 Computation time LR-CG in (s) versus problem size N for various inner solvers of shifted linear systems within CFADI preconditioning

kernels with the multicore architecture. In a multicore processor the processes have to share the cache if more than one thread is assigned to a socket. We believe that this might be an explanation for the numerical observations. Although the multithreaded parallel direct solver PARDISO improves the numerical solution of the LR-CG method with CFADI preconditioning, for larger sizes N the multilevel ILU is still superior although not yet being parallelized. This can be concluded from the comparison in Fig. 5.13 for $\text{tol}_w = 10^{-1}$ and $\text{tol}_w = 10^{-2}$.

5.5 Conclusions

In this article we have demonstrated the benefits of low-rank Krylov subspace methods. When computing the approximate solution of generalized, projected Lyapunov equations, these novel low-rank Krylov subspace comprise the benefits of Krylov subspace methods and the low-rank Cholesky factor representation similar to LRCF-ADI methods. While the superiority of low-rank Krylov subspace methods is not always confirmed in the numerical experiments, their high potential has been illustrated. We have also shown that techniques of early compressing the rank to the desired accuracy is beneficial for low-rank Krylov subspace methods. The results have demonstrated the applicability in model order reduction techniques, in particular for those problems arising from circuit simulation. We have further outlined the wide range of their usage for other problems such as parabolic partial differential equations. We believe that this numerical case study helps understanding when and how low-rank Krylov subspace methods can be used as a technique for model order reduction.

Acknowledgements The work reported in this paper was supported by the German Federal Ministry of Education and Research (BMBF), grant no. 03BOPAE4. Responsibility for the contents of this publication rests with the authors.

References

1. Amestoy, P., Davis, T.A., Duff, I.S.: An approximate minimum degree ordering algorithm. SIAM J. Matrix Anal. Appl. **17**(4), 886–905 (1996)
2. Bartels, R., Stewart, G.: Solution of the matrix equation $AX + XB = C$. Commun. ACM **15**(9), 820–826 (1972)
3. Benner, P.: Advances in balancing-related model reduction for circuit simulation. In: Roos, J., Costa, L.R.J. (eds.) Scientific Computing in Electrical Engineering SCEE 2008. Mathematics in Industry, vol. 14, pp. 469–482. Springer, Berlin/Heidelberg (2010)
4. Benner, P., Li, J.R., Penzl, T.: Numerical solution of large-scale Lyapunov equations, Riccati equations, and linear-quadratic optimal control problems. Numer. Linear Algebra Appl. **15**(9), 755–777 (2008)
5. Bollhöfer, M., Saad, Y.: Multilevel preconditioners constructed from inverse–based ILUs. SIAM J. Sci. Comput. **27**(5), 1627–1650 (2006)
6. Brenan, K.E., Campbell, S.L., Petzold, L.R.: The Numerical Solution of Initial-Value Problems in Differential-Algebraic Equations. Classics in Applied Mathematics, vol. 14. SIAM, Philadelphia (1996)
7. Damm, T.: Direct methods and ADI preconditioned Krylov subspace methods for generalized Lyapunov equations. Numer. Linear Algebra Appl. **15**(9), 853–871 (2008)
8. Drmač, Z., Bujanović, Z.: On the failure of rank-revealing QR factorization software – a case study. ACM Trans. Math. Softw. **35**(2), 12:1–12:28 (2008)
9. Eppler, A.K., Bollhöfer, M.: An alternative way of solving large Lyapunov equations. Proc. Appl. Math. Mech. **10**(1), 547–548 (2010)

10. Eppler, A.K., Bollhöfer, M.: Structure-preserving GMRES methods for solving large Lyapunov equations. In: Günther, M., Bartel, A., Brunk, M., Schoeps, S., Striebel, M. (eds.) Progress in Industrial Mathematics at ECMI 2010. Mathematics in Industry, vol. 17, pp. 131–136. Springer, Berlin (2012)

11. Freund, R.W.: SPRIM: structure-preserving reduced-order interconnect macromodeling. In: Proceedings of the International Conference on Computer Aided Design (ICCAD), pp. 80–87. IEEE Computer Society, San Jose (2004)

12. Freund, R., Jarre, F.: A QMR–based interior–point algorithm for solving linear programs. Math. Program. Ser. B **76**(1), 183–210 (1997)

13. Freund, R., Nachtigal, N.: QMR: a quasi-minimal residual method for non-hermitian linear systems. Numer. Math. **60**, 315–339 (1991)

14. Golub, G.H., Van Loan, C.F.: Matrix Computations. Johns Hopkins Studies in Mathematical Sciences, 3rd edn. The Johns Hopkins University Press, Baltimore (1996)

15. Hackbusch, W.: Hierarchische Matrizen. Springer, Berlin/Heidelberg (2009)

16. Hinze, M., Volkwein, S.: Proper orthogonal decomposition surrogate models for nonlinear dynamical systems: error estimates and suboptimal control. In: Dimension Reduction of Large-Scale Systems. Lecture Notes in Computational Sience and Engineering, vol. 45, Chap. 10, pp. 261–306. Springer, Berlin/Heidelberg (2005)

17. Hochbruck, M., Starke, G.: Preconditioned krylov subspace methods for lyapunov matrix equations. SIAM J. Matrix Anal. Appl. **16**(1), 156–171 (1995)

18. Ionuţiu, R., Rommes, J., Schilders, W.H.A.: SparseRC: sparsity preserving model reduction for RC circuits with many terminals. IEEE Trans. Comput. Aided Des. Integr. Circuits Syst. **30**(12), 1828–1841 (2011)

19. Jaimoukha, I., Kasenally, E.: Krylov subspace methods for solving large Lyapunov equations. SIAM J. Numer. Anal. **31**(1), 227–251 (1994)

20. Jbilou, K.: Block Krylov subspace methods for large continuous-time algebraic Riccati equations. Numer. Algorithms **34**, 339–353 (2003)

21. Jbilou, K.: An Arnoldi based algorithm for large algebraic Riccati equations. Appl. Math. Lett. **19**(5), 437–444 (2006)

22. Jbilou, K.: ADI preconditioned Krylov methods for large Lyapunov matrix equations. Linear Algebra Appl. **432**(10), 2473–2485 (2010)

23. Jbilou, K., Riquet, A.: Projection methods for large Lyapunov matrix equations. Linear Algebra Appl. **415**(2–3), 344–358 (2006)

24. Kressner, D., Tobler, C.: Krylov subspace methods for linear systems with tensor product structure. SIAM J. Matrix Anal. Appl. **31**(4), 1688–1714 (2010)

25. Kressner, D., Plešinger, M., Tobler, C.: A preconditioned low-rank CG method for parameter-dependent Lyapunov matrix equations. Technical report, EPFL (2012)

26. Kunisch, K., Volkwein, S.: Galerkin proper orthogonal decomposition methods for parabolic systems. Numer. Math. **90**, 117–148 (2001)

27. Levenberg, N., Reichel, L.: A generalized ADI iterative method. Numer. Math. **66**, 215–233 (1993)

28. Li, J.R., White, J.: Low rank solution of lyapunov equations. SIAM J. Matrix Anal. Appl. **24**(1), 260–280 (2002)

29. Mikkelsen, C.C.K.: Numerical methods for large lyapunov equations. Ph.D. thesis, Purdue University (2009)

30. Odabasioglu, A., Celik, M., Pileggi, L.T.: PRIMA: passive reduced-order interconnect macro-modeling algorithm. IEEE Trans. Circuits Syst. **17**(8), 645–654 (1998)

31. Penzl, T.: Lyapack — a MATLAB toolbox for large Lyapunov and Riccati equations, model reduction problems, and linear-quadratic optimal control problems (2000). Release 1.8 available at http://www.tu-chemnitz.de/mathematik/industrie_technik/downloads/lyapack-1.8.tar.gz

32. Penzl, T.: A cyclic low-rank smith method for large sparse Lyapunov equations. SIAM J. Sci. Comput. **21**(4), 1401–1418 (2000)

33. Reis, T., Stykel, T.: PABTEC: passivity-preserving balanced truncation for electrical circuits. IEEE Trans. Comput. Aided Des. Integr. Circuits Syst. **29**(9), 1354–1367 (2010)
34. Reis, T., Stykel, T.: Positive real and bounded real balancing for model reduction of descriptor systems. Int. J. Control **83**(1), 74–88 (2010)
35. Saad, Y.: A flexible inner-outer preconditioned GMRES algorithm. SIAM J. Sci. Comput. **14**(2), 461–469 (1993)
36. Saad, Y.: Iterative Methods for Sparse Linear Systems, 2nd edn. SIAM, Philadelphia (2003)
37. Saad, Y., Schultz, M.: GMRES: a generalized minimal residual algorithm for solving nonsymmetric linear systems. SIAM J. Sci. Stat. Comput. **7**(3), 856–869 (1986)
38. Sabino, J.: Solution of large-scale Lyapunov equations via the block modified smith method. Ph.D. thesis, Rice University, Houston, TX (2006)
39. Schenk, O., Gärtner, K.: Solving unsymmetric sparse systems of linear equations with PARDISO. J. Futur. Gener. Comput. Syst. **20**(3), 475–487 (2004)
40. Schenk, O., Gärtner, K.: On fast factorization pivoting methods for symmetric indefinite systems. Electron. Trans. Numer. Anal. **23**(1), 158–179 (2006)
41. Simoncini, V.: A new iterative method for solving large-scale lyapunov matrix equations. SIAM J. Sci. Comput. **29**(3), 1268–1288 (2007)
42. Starke, G.: Optimal alternating direction implicit parameters for nonsymmetric systems of linear equations. SIAM J. Numer. Anal. **28**(5), 1432–1445 (1991)
43. Starke, G.: Fejér-Walsh points for rational functions and their use in the ADI iterative method. J. Comput. Appl. Math. **46**, 129–141 (1993)
44. Stykel, T.: Low-rank iterative methods for projected generalized Lyapunov equations. Electron. Trans. Numer. Anal. **30**, 187–202 (2008)
45. Stykel, T., Reis, T.: Passivity-preserving balanced truncation model reduction of circuit equations. In: Roos, J., Costa, L. (eds.) Scientific Computing in Electrical Engineering SCEE 2008. Mathematics in Industry, vol. 14, pp. 483–490. Springer, Berlin/Heidelberg (2010)
46. Van der Vorst, H.A.: Bi-CGSTAB: a fast and smoothly converging variant of Bi-CG for the solution of nonsymmetric linear systems. SIAM J. Sci. Stat. Comput. **13**(2), 631–644 (1992)
47. Wachspress, E.: Iterative solution of the Lyapunov matrix equation. Appl. Math. Lett. **107**, 87–90 (1988)

Index

$\mathscr{H}_\infty$-norm, 46

alternating direction implicit, 67
alternating direction implicit method, 160
Analog Insydes, 136, 137, 144
asymptotically stable, 47
augmented matrix, 121, 122

balanced, 49
balanced truncation, 48, 116, 158
  introduction, 158
  numerical results, 173
BBD. *See* bordered block diagonal
block moment, 92
bordered block diagonal, 106
boundary conditions, 5, 8, 9, 14
branch constitutive relations, 42
build-in potential, 6

characteristic values, 49
circuit simulation, 87, 89
conductance matrix, 89
conductivity, 95
contact, 5, 28
controllability Gramian, 49
correction equation, 124, 125
coupled system, 24, 28
coupling condition, 5, 8, 12
current density, 4, 5
current sources, 3, 28
cutset, 41
CV-loops, 44

DAE. *See* differential-algebraic equation
DASPK, 12, 31
DD equation. *See* drift-diffusion equation
DEIM. *See* discrete empirical interpolation
  method
descriptor system. *See* system, descriptor
  reciprocal, 114
dielectricity, 4, 5
differential-algebraic equation, 2, 3, 5, 8–12,
  16, 24, 28–31, 90, 135, 137
differentiation index, 44
directed graph, 41
discrete empirical interpolation method, 3,
  19–23
discretization, 2, 7–12
doping profile, 4
double orthogonal, 123
drift-diffusion equation, 2, 4, 6, 8

(E)SVDMOR. *See* (extended) singular value
  decomposition MOR
(extended) singular value decomposition
  MOR, 88, 96–100, 108–120
electrical network, 2–7, 17–18, 28
electro-quasistatic, 95
electron and hole concentration, 4, 9
electronic circuits
  differential-amplifier circuit, 146
  operational amplifier circuit, 150
  transmission line, 146, 149
electrostatic potential, 4
elementary charge, 4
EMMP. *See* extended moment matching
  projection

© Springer International Publishing AG 2017
P. Benner (ed.), *System Reduction for Nanoscale IC Design*,
Mathematics in Industry 20, DOI 10.1007/978-3-319-07236-4

Printed in the United States
By Bookmasters